TRIAL
BY
JURY

TRIAL
BY
JURY

KEVIN KING

DESIGNED AND ILLUSTRATED
BY

Austin Stevens

NEWBURY HOUSE PUBLISHERS, INC.
ROWLEY, MASSACHUSETTS 01969
ROWLEY • LONDON • TOKYO

1 9 8 4

Library of Congress Cataloging in Publication Data

King, Kevin.
   Trial by jury.

   1. Trials--United States.   2. Law--United States--
Terms and phrases.   3. English language--
Text-books for foreigners.   I. Title.
KF9655.Z9K56   1984   345.73'02   83-21948
ISBN 0-88377-370-8   347.3052

NEWBURY HOUSE PUBLISHERS, INC.

Language Science
Language Teaching
Language Learning

ROWLEY, MASSACHUSETTS 01969
ROWLEY • LONDON • TOKYO

Printed in the U.S.A.

First printing: February 1984

86 87 88 89   10 9 8 7 6 5 4 3

CONTENTS

The author wishes to thank Anne Dow and Bill Biddle, Harvard University, for their help in the preparation of this book. Gratitude is also expressed to Chris Braider, Barry Megquier, Ed Shloth, Margaret Sullivan and Beth Wellington for assistance with the tapes.

HOW TO USE *TRIAL BY JURY*

This book is primarily a discussion text, but it is also designed to improve reading and listening skills. It involves the intermediate ESL/EFL student in problem-solving activities and vocabulary work pertaining to twelve court cases. In each case, the student goes from reading the basic situation all the way to acting out the resulting trial in the classroom. Between the first reading and the final trial, the student works on vocabulary and comprehension exercises in addition to other kinds of practice work related to the case. The student also gives opinions about the legal aspects of the case and listens to taped conversations between characters involved in the events leading up to the trial. Multiple-choice listening comprehension questions follow, along with other post-tape activities. These help the student isolate the main issues to be considered in the trial. The work can be done entirely in class, but if discussion is the teacher's foremost goal, as much of the work as possible should be done at home before class.

PREPARED AT HOME BEFORE CLASS

A The *vocabulary* section. After the student has read the case, he/she works on this self-teaching exercise. The teacher must emphasize that the best definition for each word will be found from the context. For this reason the number of the paragraph in which the word appears is given.

B The *cumulative vocabulary exercise* section. This tests the student's recall of the words he/she has learned in previous lessons or in the legal lexicon (Appendix 1). For example, "The j_ _ _ _ _ _ wondered what the j_ _ _ _ _ would think about the case." Looking at the legal lexicon, the student finds that, of words beginning with "j," *judge* is correct for the first blank, and *jury* is correct for the second blank.

C The *general exercises* section. There are various types of exercises, including "sequencing," in which events are listed out of order. The student puts a number next to the event to indicate its proper place in the sequence of events. In the "fill in with the correct fact" exercises, the student should try to fill in the fact from memory. If he/she cannot, the student should refer to the text.

D The *law questions* section. After each question there are three boxes:

Your decision	*Jury's decision*	*Score*
☐	☐	☐

The student, at home, fills in *only* the first box ("Your decision") with his/her choice of a, b, c, etc.

E The *pre-trial questions* section. Given a quote, the student determines who the speaker is, or if the statement would be used by the prosecution or by the defense. In addition, the student may be asked to write several questions which he/she would ask one or more of the participants in the trial.

WORK TO BE DONE IN CLASS

The self-teaching vocabulary exercise, the cumulative vocabulary, and the general exercises may be corrected by the teacher—who merely reads the right answer—in a matter of a couple of minutes. However, I find it preferable for the students to correct them themselves in small groups (3–5) while the teacher oversees the whole process. I've found that most students usually have the right answer. This process will take less than five minutes and will encourage self-reliance and free speaking. The one exception is the paragraphing exercise, which must be examined by the teacher.

While in small groups the students compare their answers to the law questions. The teacher must emphasize that there is no *right* answer. Each group is in and of itself a jury, and the majority opinion rules. The students discuss the question in order to convince dissenters of their errors. A vote is taken and the majority of the jury's decision is recorded in the box thus designated. If a majority agreement cannot be reached, nothing is recorded in that box (in which case everybody in that group gets a zero in the *Score* box). In this case the teacher should prod the students. If they seem hopelessly deadlocked (two a's and two b's; or in the case of three students—one with a, one with b, and one with d), the teacher should call for a vote. If no one changes his or her mind, they all get a zero and must go to the next question so as not to waste time.

If a discussion group finds none of the given answers (a, b, c, d, e) appropriate and agrees unanimously on a different answer, this new answer (e.g., "f") should be written into their books. The letter "f" will be put in the box marked *Jury's decision*, and a "2" will be put in the box marked *Score*. (See the explanation of scoring below.) The score should be recorded immediately after the discussion. The scoring system was initiated to encourage students to argue for their points of view, rather than just saying them and going on to the next question.

Scoring. If the student's answer is the *same* as the majority of the jury's decision, the student gets a "2."

If the student's answer is *different* from the majority decision, but the student changes his/her mind and agrees with them, the student gets a "1."

If the student's answer is different from the majority of the jury's decision and the student does not change his/her mind, the student gets a "0."

In short form: same—2
different & change—1
different & no change—0
no majority—0

Remember—there is *no right answer!* Each group is a jury unto itself. Each member of the jury must try hard to make the other jurors agree with his/her decision in order to get the highest score. After a few questions the system becomes automatic. Students make no errors with it.

Let's take a hypothetical example to illustrate how it works. A jury consists of three students named Jules, Jim, and Jill. They discuss Law Question 5 of Case 1, which is on page 2. Jules and Jim both think that they would *shoot him* (a). Jill had decided at home that she would *try to run away* (b). After a few minutes of discussion, Jill is convinced that Jules and Jim are right. She changes her mind and agrees with them. Their books would be marked thus:

	Your decision	Jury's decision	Score
Jules:	a	a	2
Jim:	a	a	2
Jill:	☒ a	a	1

The questions are designed so that more than one opinion is tenable. Sometimes one entire group will agree on an answer, and another group will unanimously agree on a different answer for the same question.

The teacher should avoid giving opinions. The teacher's role during the discussion of the law questions is listening, correcting speech errors, and making sure that a group doesn't get bogged down on any question. The teacher should control the pace, encouraging slow groups to speed up, sometimes insisting on a vote and going on to the next question.

The teacher should emphasize peer correction. When the teacher hears an obvious error, he/she may stop the speaker and ask the others what it was. Generally, it will have been noticed. The teacher may then tell the correcting student that he/she or anyone else in the group must do the same thing when the teacher is listening to a different group. Peer correction is an essential part of the discussion format, and must be repeatedly emphasized.

F The *listening comprehension* section. The students listen to a tape which features a conversation between two or more of the characters in the case. The conversation is followed by questions. The students circle their choices of a, b, or c as the correct answer. The teacher then gives the right answer and responds to any questions the students have about the tape.

New and essential information regarding the guilt, innocence, or litigation stance of the characters is provided. The tapescripts are included in Appendix 2 so that in some cases the characters' parts can be taken by students or read by the teacher.

G The *post-tape questions* section. Armed with the information learned from the tape, the students can write several questions—sometimes incisive, poignant, or damning—to be asked of the characters during the trial. These may be corrected as they are being written or, if class time has run out, can be handed in the next day, after the trial. It is important for the teacher not to judge the shrewdness of the questions, but to concentrate on grammar, spelling, and so on. It is enough that students generate good, simple questions which will extract information from the witnesses and defendants. (For example, in Case 1 a student might write the following, directed at the gun store clerk: "Why did you sell the gun to Wolfgang?")

H *The trial.* Before the trial actually begins, the teacher should invite questions from the class. Not all the details can be covered in a few pages. Students may ask, for example, "Do they have insurance?" (I find "no" to be the most effective answer to this particular question.) Any decision on questions such as this can work, as long as everyone is working with the same assumptions. Similarly, the teacher may wish to make his/her own comments on the case at this point (for example, making sure that students know the great difference between *drinking* and being *drunk*).

The major participants should sit in front of the class with antagonists on opposite sides of the judge (teacher). The jurors should sit facing the participants, as close as possible. For example, in Case 1 the class/courtroom would look like this:

Judge

Wolfgang	(desk)	Johannes	Clerk

Jurors

The teacher explains that this trial is specifically for an ESL classroom. Therefore, all the class members who do not play roles as defendants or witnesses are the jurors who will decide the case. *They must pretend that they know nothing about the case!* They must ask questions to find out what happened. The teacher must be careful *not* to allow a question like "*Why* did you hang up the phone?" to occur before the question "*Did* you hang up the phone?" has been asked.

The jurors should refer to their pre-trial and post-tape questions to help formulate questions, but this should play a small part in the questioning process.

The teacher is the judge, and he/she designates who will speak. The role of the judge is to encourage jurors who haven't asked questions to do so, and to pace the trial, intervening when a witness becomes repetitious or when the questions aren't leading anywhere. Lying on the part of the defendant is permissible, but needn't be encouraged since it will come naturally to any player in a tight spot. It often produces humorous situations. The teacher may want to experiment with a student as the judge at some point. The advantages of this are that it frees the teacher to concentrate more on the errors being made, and the students feel more independent.

Whenever the judge thinks a minor witness (bartender, child, doctor, etc.) could clear up some confusion or could contradict a player who has grossly lied, one of the jurors should be appointed to take the stand as that minor player, and then may resume his/her role as juror after testifying and being questioned.

The judge should decide when all important issues have been discussed. The trial should not drag. In general, trials last about a half hour. The teacher declares the trial finished and asks the jury to convene (in a circle if possible). The teacher should tell the jury the issue(s) it has to decide on (as few and as simple as possible), and should make sure the discussion is channeled toward these points. The students who participated as players in the trial remain where they are, awaiting the verdict.

The teacher may want to keep tighter control on the jury's deliberation than on the trial, since it is easy to stray from the issues that must be decided. (In Case 1, for example, the teacher might say, "Did Wolfgang murder Johannes or was it an accident?") The deliberation should be relatively swift (about five minutes). It often works well to ask the jury chairperson to ask for a vote. (In Case 1 he/she might say, "Raise your hand if you think Wolfgang is guilty of murder." If the verdict is guilty [majority decision], the teacher should ask, "Should Wolfgang go to jail?" If the majority raise their hands for "yes," the teacher then asks, "For how long?" Each person in turn says how long, and the chairperson figures the average. He/she then announces the sentence.)

Another possibility is to have each student tell, in a few sentences, his/her opinion. A good student should be appointed jury chairperson. The chairperson has the duty of announcing the verdict—guilt or innocence—the sentence, or how much will be paid by whom for damages.

If the trial proves exciting, the jury deliberation can easily be put off till the next day. The decision (verdict) can be made in a few minutes after a short refresher of the facts by the teacher.

The *Result* is then written in the designated place in the book.

TIME

The case time span will vary from class to class. Much depends on the teacher's pacing of the discussion of the law questions, the trial, and the jury's deliberation. Each teacher will have to establish what works best. All cases can be completed in an hour and a half. If there is not time for a second class, the trial may be omitted. In a fifty-minute class, the average time breakdown will be something like this:

(1st class)
vocabulary and exercise correction	5 min.
law question discussion	30 min.
listening comprehension	15 min.

(2nd class)
trial	30 min.
jury deliberation	10 min.
	90 min.

OPTIONAL USES

Trial by Jury may be used in any class for additional practice in (1) listening comprehension, (2) writing, and (3) discussion. The teacher may assign the self-teaching vocabulary exercise and the law questions for homework. The students will do the listening comprehension section in class the next day, and they will write their opinions of the best resolution of the case that night. The teacher may want to assign the post-tape questions as an in-class writing assignment or as additional homework.

Trial by Jury also can be used profitably by advanced classes, with teachers encouraging the use of more advanced structures and vocabulary, omitting the present vocabulary and general exercises sections. These classes might be broken into two groups, A and B, doing different cases simultaneously. The advantage of this is that when group A goes to trial, the members of group B become the jury, and vice versa. Jury members then need not *pretend* unfamiliarity with the case; they will have neither read nor discussed it.

The teacher is the judge, and he/she designates who will speak.

TRIAL
BY
JURY

. . . he twisted Wolfgang's hand badly.

THE CASE OF
THE PIANO TEACHER

(1)   Wolfgang was a piano teacher who came from Germany to New York seven years ago. He was very poor and had a student named Ludwig who was also very poor. Ludwig was always late in paying Wolfgang for his lessons. Wolfgang was a thin man and was not very strong, but he would hit young Ludwig, age 13, when he was late in paying. Wolfgang could not control his behavior when he got mad.

(2)   One afternoon, Ludwig went home with a black eye. Ludwig had an older brother, Johannes, who was cruel and had a terrible temper. When Johannes saw what had happened to his younger brother, he went to Wolfgang's apartment, hit him a few times, and twisted his hand badly. Johannes told Wolfgang that if he ever hit young Ludwig again, he would break both of his hands. Wolfgang was frightened. He had no doubt that Johannes had both the ability and the desire to do so.

(3)   Wolfgang spoke bad English. He thought that Johannes had also said he might kill him. Wolfgang was so afraid that he went out and bought a gun as soon as he could afford one.

(4)   The next week Ludwig again could not pay on time. Wolfgang was furious. He had no money. He had spent it all on the gun. He hit Ludwig hard and knocked out a tooth.

(5)   When Johannes found out, he took Ludwig to Wolfgang's apartment. Wolfgang told them to go away, but Johannes forced the door open. He picked up a chair and said he was going to break Wolfgang's hands so that he would never play the piano again. Wolfgang pulled the gun from his pocket and fired. He missed. Johannes and Ludwig ran out. Wolfgang fired again, this time hitting Ludwig and killing him. Wolfgang says that he fired twice with almost no time between shots, but Johannes says that at least three seconds passed between the shots.

A. VOCABULARY

Circle the letter next to the answer that is similar to the word(s) on the left.

1. temper   a. finger
   (para. 2)   b. book
                c. grade
                (d.) behavior
                e. hand

2. twisted   a. kissed
   (para. 2)   (b.) turned
                c. shook
                d. looked at
                e. touched

3. furious   a. one-eyed
   (para. 4)   b. happy
                (c.) angry
                d. dead
                e. alive

4. fired   (a.) shot
   (para. 5)   b. smiled
                c. sat down
                d. fell
                e. burned

B. CUMULATIVE VOCABULARY EXERCISE

Fill in with words from this case and from the legal lexicon (Appendix 1).

The j _ _ _ _ e wondered what the j _ _ _ _

would think of this case. Would the v _ _ _ _ _ _ _

be guilty or innocent? He himself often got

f _ _ _ _ _ _ _ when he wasn't paid on time.

1

Sometimes he wanted to t _ _ _ _ the neck of the city clerk. But he had learned to control his t _ _ _ _ _ _ . He never did anything i l _ _ _ _ _ _ .

C. GENERAL EXERCISES

1. *Sequencing:* Put a number (1–6) in the blank to denote the order in which the events happened. Try not to look back at the story.

a. ____ Wolfgang shot Ludwig.

b. ____ Wolfgang bought a gun.

c. ____ Wolfgang knocked out Ludwig's tooth.

d. ____ Johannes picked up a chair.

e. ____ Ludwig got a black eye.

f. ____ Johannes twisted Wolfgang's hand.

2. *Fill in with the Correct Fact:* (Do as many as possible from memory.)

   1. Johannes claims there were at least _____ seconds between shots.

   2. Ludwig was _____ years old.

   3. In total, Wolfgang fired _____ shots.

   4. Wolfgang carried his gun in his _____ .

D. LAW QUESTIONS

Answer the questions. Put the letter that corresponds to the best answer in the box marked *Your decision*.

1. Wolfgang had a bad temper. This is a

   a. good defense because we should forgive him.
   b. good defense because everyone has a bad temper once in a while.
   c. bad defense because he is still responsible for his actions.
   d. bad defense because he is poor.

   Your decision   *Jury's decision*   *Score*
   ☐   ☐   ☐

2. For Wolfgang, getting his hands broken is

   a. more serious than for most people because he plays the piano.
   b. more serious than for most people because he needs his hands to defend himself.
   c. less serious than for most people because he can't write English.

   d. less serious than for most people because he is poor.
   e. more serious than for most people because he needs strong hands to hit his students.

   ☐   ☐   ☐

3. Did Wolfgang really need a gun?

   a. Yes.
   b. No.

   ☐   ☐   ☐

4. Wolfgang and Johannes live in a big modern city. If ordinary people (not police officers or soldiers) could not own guns, this city would be

   a. safer.
   b. less safe.

   ☐   ☐   ☐

5. If you were Wolfgang, and Johannes broke down the door and said he was going to break your hands, you would

   a. shoot him.
   b. try to run away, and call the police later.
   c. tell Johannes to sit down and discuss it.
   d. shout "Help!"

   ☐   ☐   ☐

6. The amount of time between the shots is important because

   a. if there were three seconds between shots, then Wolfgang wanted to kill someone. It was not self-defense.
   b. time is money.
   c. it takes three seconds to shout "Help!"
   d. if there was almost no time between shots, then Wolfgang was shooting in self-defense.
   e. both *a* and *d*.

   ☐   ☐   ☐

7. Wolfgang killed young Ludwig instead of Johannes. If Wolfgang had killed Johannes,

   a. he should go to jail.
   b. he should go free (not to jail).
   c. he should go free (not to jail) only if he shot in self-defense.

   ☐   ☐   ☐

8. Poor little Ludwig is dead. Which of these arguments is not a good defense for Wolfgang?

   a. It was an accident.
   b. Wolfgang shot in self-defense.

c. Johannes is to blame for breaking down the door and bringing Ludwig with him to Wolfgang's house.
d. Ludwig didn't pay on time.

☐   ☐   ☐

9. What should happen to Wolfgang?

   a. He should go free.
   b. He should go to jail for 1 year.
   c. He should go to jail for 5 years.
   d. He should go to jail for 10 years.
   e. The judge should break his hands.

☐   ☐   ☐

10. Is it important that Wolfgang spoke very bad English and could not understand it well?

   a. Yes.
   b. No.

☐   ☐   ☐

11. One famous lawyer said, "Society is to blame for this death." What did the lawyer mean?

   a. The United States does not provide enough English lessons for people who come to live here.
   b. U.S. society makes it easy for anyone, even a madman, to get a gun.
   c. It is very difficult for a musician to make enough money to live on in the United States.
   d. It is difficult for a foreigner to get a job in the United States.
   e. All of the above.

☐   ☐   ☐

Total Score _____

E. PRE-TRIAL QUESTIONS

Who is likely to make each argument? In the blanks write *P* for the *prosecution* or *D* for the *defense* of Wolfgang.

1. ____ It was an accident.

2. ____ It was self-defense.

3. ____ Society is also guilty.

4. ____ Johannes is partly guilty.

5. ____ Wolfgang shouldn't be so worried about his fingers. There are other jobs besides teaching piano.

6. ____ Wolfgang did not understand Johannes because his English was not good.

7. ____ Two shots were fired, one immediately after the other.

8. ____ Three seconds passed between shots.

9. ____ Wolfgang bought a gun because he was planning to kill Johannes.

F. LISTENING COMPREHENSION

Listen to the tape, then circle the correct answer.

1.   a   b   c
2.   a   b   c
3.   a   b   c
4.   a   b   c
5.   a   b   c
6.   a   b   c
7.   a   b   c   d

G. POST-TAPE QUESTIONS

After listening to the tape, write three questions that you, as a juror, would ask the clerk at the trial.

1. _____

   _____

2. _____

   _____

3. _____

   _____

H. THE TRIAL

Defendant: Wolfgang.
Witnesses: Johannes; the gun store clerk.
Charge: The murder of Ludwig.
Verdict: In order for Wolfgang to be guilty of murder, the jury must agree (majority vote) that he intended to kill.

If the jury decides that the killing was accidental, then Wolfgang is guilty of the crime of *homicide*, which is far less serious.

If the jury decides that Wolfgang acted in self-defense, then he is innocent of all charges.

Sentence: Murder—life imprisonment or death. The jury votes on which to impose.

Homicide—the jury votes on how many years or months in prison.

Procedure: The judge announces that Wolfgang is on trial for the murder of Ludwig and asks any juror to start the questioning of any of the witnesses.

Result: _____

   _____

   _____

   _____

   _____

3

THE CASE OF LINKS' RIGHTS

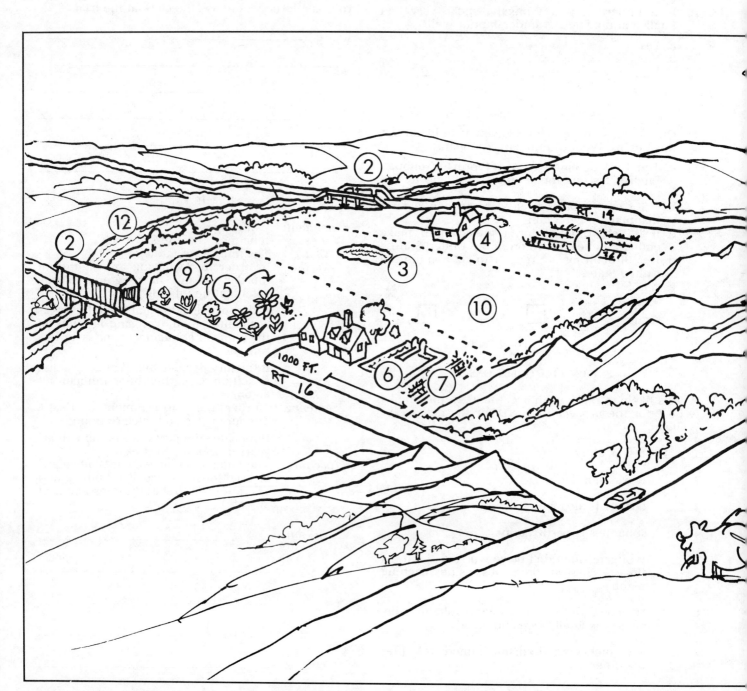

(1)   Farmer Brown lived in a big house on Route 16. The animals were eating all his corn, so he decided to move to the city. He sold his land to Mr. Zitkopf. Farmer Brown wrote the following contract, and they both signed it: "I sell my land on Rt. 16 to Mr. Zitkopf. The land is 1000 ft. long, from the mountains in the west to the river in the east. It is 400 ft. deep, ending at Mr. Links' land. Mr. Links has a right of way through my land to his."

(2)   Mr. Zitkopf knew that no one had seen Mr. Links for twenty years, and everyone assumed that he was dead. So Zitkopf put in a swimming pool on the east side of his house and a flower garden on the west side. He grew lettuce and tomatoes in Brown's old vegetable garden next to the swimming pool.

(3)   One Sunday he found that someone had driven across his lettuce! He soon discovered it was Mr. Links, who was fishing in his pond.

(4)   Zitkopf was furious. "Everyone knows you're dead!" he said to Links. "You must be a ghost."

"I'm no ghost," laughed Links, "and I have the right of way to drive across your land to mine."

"But you drove right through my vegetable garden!" said Zitkopf. "What kind of neighbor are you?"

"I'm not stupid," said Links. "I couldn't drive across your swimming pool, so I had to choose between the flowers and the lettuce. I like flowers, but not lettuce."

"You could have used the path on the other side of the trees by the river," said Zitkopf.

"There are a lot of sharp rocks there, as well as holes and broken bottles. I'm not going to ruin my tires," said Links.

"You're going to pay for the lettuce you ruined! I'll see you in court!" said Zitkopf.

"I have the right of way," said Links as he left.

The animals were eating all his corn.

THE VIEW FROM THE SOUTHEAST

1. Joe's garden

2. bridges

3. Links' pond

4. Joe's land

5. Zitkopf's flower garden

6. Zitkopf's swimming pool

7. Zitkopf's vegetable garden

8. mountains

9. rocky path

10. Links' land

11. state land

12. river

A. VOCABULARY

Circle the letter next to the answer that is similar to the word(s) on the left.

1. contract
   (para. 1)
   a. novel
   b. poem
   c. recipe
   d. legal agreement
   e. book

2. right of way
   (para. 1)
   a. correct method
   b. correct number of points
   c. permission to pass
   d. not wrong
   e. not on the left

3. ghost
   (para. 4)
   a. person who invites
   b. person who is invited
   c. relative
   d. unreal or dead form that looks alive
   e. person who breaks the law

4. ruin
   (para. 4)
   a. buy
   b. sell
   c. help
   d. place to buy or look at old things
   e. destroy or damage

B. CUMULATIVE VOCABULARY EXERCISE

Fill in with words from this case, the previous case, and the legal lexicon (Appendix 1).

Two years ago, Farmer Brown saw someone stealing lettuce from his garden at midnight. The

t _ _ _ _ _ ran away when he saw Brown lift his

gun. Brown f _ _ _ _ _ from twenty feet and

missed. "How could I miss," said Brown. "He must have

been a g _ _ _ _ _ ." Joe was with him. He

w _ _ _ _ _ s _ _ _ _ the whole thing. Brown's

bullet missed the t _ h _ _ _ _ , but it r _ u _ _ _ _ _

a new rose bush.

C. GENERAL EXERCISES

Identify: Look at the map. Put the correct letter next to the objects listed below. Refer to the text but not to the other map.

1. Zitkopf's swimming pool _____

2. Joe's land _____

3. river _____

4. mountains _____

5. bridges _____

6. Joe's garden _____

7. Zitkopf's vegetable garden _____

8. state land _____

9. Links' land _____

10. Zitkopf's flower garden _____

11. Links' pond _____

12. rocky path _____

D. LAW QUESTIONS

Answer the questions. Put the letter that corresponds to the best answer in the box marked *Your decision.*

ADDITIONAL FACTS FOR THE COURT:

1. The state *must* provide a road across state land for any person who has no other way to get to his or her land. The mountains on state land just east of Links' land are difficult to pass through.
2. Mr. Links is an excellent swimmer.

1. What *mistake* did Brown make in writing the agreement to sell land to Zitkopf?

    a. He doesn't really live on Route 16, but on the river.
    b. His land starts—not ends—at Links' land.
    c. He forgot to mention the pond.
    d. The river is in the north.
    e. The mountains are in the east, and the river is in the west, but Brown reversed these.

Your decision   *Jury's decision*   *Score*

☐   ☐   ☐

2. Because of this mistake,

    a. Links doesn't own any land.
    b. Links doesn't have a right of way.
    c. Brown still owns the land he had sold.
    d. the mountains must be moved.
    e. nothing should happen: the mistake is unimportant.

☐   ☐   ☐

3. Links is an excellent swimmer. This fact is

    a. relevant because he can have his right of way across the river.
    b. irrelevant because the water freezes in the winter.
    c. irrelevant because the ice is too thin to walk on.
    d. irrelevant because Links still has a right of way across Zitkopf's land.
    e. irrelevant because Links doesn't have a bathing suit.

☐   ☐   ☐

4. Zitkopf says that he doesn't have to give Links a right of way because he thought Links was dead. This is

    a. reasonable.
    b. unreasonable.
    c. reasonable if the swimming pool freezes.
    d. reasonable unless Brown dies.
    e. unreasonable because Brown might die.

☐   ☐   ☐

5. In this case, the people in this town have

    a. a special duty to help Links.
    b. a special duty to help Zitkopf.
    c. a special duty to help Joe.
    d. the same duty to help everybody.
    e. a special duty to help students.

☐   ☐   ☐

6. Can Zitkopf ignore Links' right of way?

    a. Yes, because he built a swimming pool.
    b. Yes, because Links has other ways to get to his land.
    c. Yes, because individuals can do anything they want with their land.
    d. No, because he signed an agreement to let Links cross his land.
    e. Yes, because a signed agreement doesn't matter when somebody is a troublemaker.

☐   ☐   ☐

7. In driving across the vegetable garden, Links acted

    a. reasonably because he had to destroy something to get there.
    b. reasonably because flowers are more valuable than lettuce.
    c. reasonably because he doesn't like salad.
    d. unreasonably because he didn't discuss the problem with Zitkopf.
    e. unreasonably because troublemakers are always unreasonable.

☐   ☐   ☐

8. Zitkopf wants to give Links his right of way on a path by the river. This is

    a. reasonable because it is possible.
    b. reasonable because there is no other way.
    c. unreasonable now because of the holes, sharp rocks, and broken glass.
    d. unreasonable now because Links might get stuck in the snow.
    e. reasonable because Links is a good swimmer.

☐   ☐   ☐

9. Links should

    a. pay for the lettuce he destroyed.
    b. not pay for the lettuce he destroyed.
    c. pay for the flowers he destroyed.
    d. pay for the swimming pool.

☐   ☐   ☐

10. At the trial, Zitkopf mentions Links' reputation as a troublemaker. This is

    a. fair because Links obviously has not changed.
    b. fair because Links obviously came back from Canada only to make trouble for Zitkopf.
    c. fair because troublemakers always remain troublemakers.
    d. unfair because Links changed his ways when he was in jail.
    e. unfair because you really don't know if he has changed after 20 years.

☐   ☐   ☐

Total Score _____

E. PRE-TRIAL QUESTIONS

Zitkopf brings Links to court to force him to pay for the lettuce he destroyed. The judge realizes that the right-of-way problem must be solved. Which arguments are likely to be advanced by Links, and which by Zitkopf? (Write *L* or *Z* in the blanks.)

1. _____ The state should make a road through the mountains.

2. _____ Lettuce doesn't cost that much.

3. _____ If you don't use your right of way for 20 years, you lose it.

4. _____ There is a good path by the river.

5. _____ You can swim across the river.

6. _____ When I bought the land, you didn't say you would use the right of way.

7. _____ You signed the agreement. It is still good.

8. _____ You just want to make trouble.

9. _____ Sharp rocks and broken glass would ruin my tires.

You are Mr. Links. Write two questions you would ask Zitkopf at the trial.

1. _____

_____

2. _____

_____

You are Mr. Zitkopf. Write two questions you would ask Links at the trial.

1. _____

_____

2. _____

_____

F. LISTENING COMPREHENSION

Listen to the tape, then circle the correct answer.

1.   a   b   c
2.   a   b   c
3.   a   b   c
4.   a   b   c
5.   a   b   c
6.   a   b   c
7.   a   b   c
8.   a   b   c

G. POST-TAPE QUESTIONS

You are a neighbor and want the problem solved fairly. Write two questions you would ask Joe at the trial. Then write two questions you would ask Farmer Brown.

(to Joe)

1. _____

_____

2. _____

_____

(to Farmer Brown)

1. _____

_____

2. _____

_____

H. THE TRIAL

Defendants: Zitkopf; Links.

Witnesses: Brown; Joe; the state representative. They take seats next to Zitkopf.

Suits: (a.)   Links is suing Zitkopf for the right of way he lost to the swimming pool.

(b.)   Zitkopf is suing Links for the lettuce he destroyed.

Verdict: (a.)   If the judgment is for Links, then the jury must decide where the right of way shall be.

(b.)   The jury must decide how much money, if any, Links must pay Zitkopf.

Procedure: The teacher should draw the diagram of the properties on the board before the trial and encourage the players to refer to it in their answers. Any juror may begin the questioning.

Result: (a.) _____

_____

(b.) _____

_____

Mr. Zitkopf put in a swimming pool on the left side of his house.

9

THE CASE OF MICK
AND THE JUGGERNAUTS

*Mick told the old farmer
that they played loud rock and roll.*

(1) Mr. McDonald had a farm on which there were two houses. He lived in one with his wife and three children, and nobody lived in the other. Mr. McDonald was a poor farmer, and he decided to rent out the second house which was 200 yards from his house. You couldn't even seen it through the trees.

(2) At this time a famous rock and roll group called Mick and the Juggernauts was looking for a place in the country where they could practice and not bother anybody. They signed a one-year lease with Mr. McDonald. Mick told the old farmer that they played loud rock and roll. He asked McDonald if that would bother him.

(3) "No," laughed McDonald, "a little music never hurt anyone." In the next month McDonald changed his mind. Mick and the Juggernauts liked to practice from midnight till 4 A.M. The music was so loud that McDonald had trouble sleeping. His 500 chickens produced 10 percent fewer eggs. His fifty cows produced 10 percent less milk, and his kids said it tasted sour. McDonald figured his total loss to be about $400.

(4) He complained to Mick. Mick complained that the chickens woke him up at 5 A.M., and the cows would "moo" in his window.

(5) McDonald soon had another problem. Teenage girls were in love with Mick. They would come from Boston and walk across his land to listen to the group play their music. Sometimes they would sleep in the woods.

(6) One day one of these girls, Bianca, fell and broke her leg climbing a stone wall between Mick's house and McDonald's. She sued McDonald and Mick and the Juggernauts for $1,000.

(7) McDonald told the judge he no longer wanted Mick and the Juggernauts living in his house. He also wanted them to pay him for the 10 percent loss in profit on eggs and milk.

A. VOCABULARY

Circle the letter next to the answer that is similar to the word(s) on the left.

1. yard
   (para. 1)
   a. 3 feet
   b. 3 miles
   c. 3 inches
   d. 3 quarts
   e. 3 light years

2. rock and roll
   (para. 2)
   a. a kind of chair
   b. terrorist
   c. football
   d. a kind of music
   e. stone-breaking

3. lease
   (para. 2)
   a. a check for $1,000,000
   b. painting
   c. agreement to rent
   d. record
   e. place where chickens live

4. kids
   (para. 3)
   a. wives
   b. cows
   c. chickens
   d. cousins in France
   e. children

5. sour
   (para. 3)
   a. delicious
   b. hot
   c. opposite of sweet
   d. white
   e. wonderful

6. complained
   (para. 4)
   a. expressed unhappiness
   b. expressed happiness
   c. laughed
   d. went home
   e. said nothing

7. moo
   (para. 4)
   a. die
   b. dance
   c. make the sound cows make
   d. sing like a chicken
   e. lay eggs

8. teenage
   (para. 5)
   a. small
   b. small but intelligent
   c. under 13
   d. from 13 to 19 years old
   e. over 50

9. woods
   (para. 5)
   a. tops of trees
   b. place with many trees
   c. should
   d. place with snow and no trees
   e. very high mountains

10. climbing
    (para. 6)
    a. going up or over an object
    b. going under an object
    c. going around a mountain
    d. swimming
    e. eating

B. CUMULATIVE VOCABULARY EXERCISE

Fill in with words from this case, the previous cases, and the legal lexicon (Appendix 1).

Bianca and her t _ _ n _ _ _ _ friend Shirley were walking through the w _ _ _ _ on McDonald's land one night. Shirley was scared because it was i _ _ _ _ _ _ l . "What if they catch us?" said Shirley. "Do you think they would a _ _ _ _ _ t us?"

"Not for such a small c _ _ _ _ e ," said Bianca. "And if they do, we will just throw ourselves on the m _ _ _ _ y of the court."

"What do you think the s _ _ t _ _ _ _ _ would be?" asked Shirley.

Bianca laughed as she c _ _ m _ _ _ _ over a fallen tree. "The cruelest thing would be to make us listen to Mozart and Beethoven. It's such i r _ _ _ _ _ _ _ _ _ music for the 1980's," she said.

"Yes," said Shirley. "Just the thought of it leaves a s _ _ _ _ taste in my mouth."

C. GENERAL EXERCISES

Word Exchange: Look back at paragraph 5. Write a similar but new paragraph here, filling in the blanks with different words. Your new paragraph must make sense.

_____ (1) soon had another _____ (2).
   (name)   (noun)

_____ (3) girls were in love with _____ (4).
  (adj.)   (name or thing)

They would _____ (5) _____ (6) _____ (7)
            (verb)   (prep.)   (name of city)

and _____ (8) _____ (9) _____ (10).
       (verb)   (prep.) (noun or pronoun)

Sometimes _____ (11) would _____ (12)
          (pronoun)   (verb)

_____ (13) the _____ (14).
  (prep.)   (noun)

D. LAW QUESTIONS

Answer the questions. Put the letter that corresponds to the best answer in the box marked *Your decision.*

1. When Mick told McDonald about his group's music, he was acting

   a. reasonably.
   b. unreasonably.
   c. reasonably because he was lying.
   d. unreasonably because he was lying.
   e. foolishly.

   Your decision   *Jury's decision*   *Score*
   ☐   ☐   ☐

2. Did Mr. McDonald make a mistake?

   a. Yes, by not asking for enough money for the rent.
   b. Yes, by asking for too much money for the rent.
   c. Yes, by not asking when they played or finding out how loud it was.
   d. No, because he shouldn't sleep so much anyway.
   e. No, but his wife and children did.

   ☐   ☐   ☐

3. McDonald's children say the milk has turned sour. The jury should

   a. not consider this as evidence because the children aren't experts in milk-tasting.
   b. consider this as evidence because you don't have to be an expert to know when milk is sour.
   c. not consider this because it is summer.
   d. not consider this because cows sometimes eat lemons.
   e. not consider this as evidence until they taste the milk themselves.

   ☐   ☐   ☐

4. McDonald wants $400 from Mick and the Juggernauts to pay for the 10 percent loss in eggs and milk. Their best defense is:

   a. McDonald knew about the music when he signed the lease. Now the group has no responsibility.
   b. They don't have enough money to pay it.
   c. They don't like eggs or milk.
   d. The price of eggs has gone down.
   e. There might be other reasons besides the music for the lower production of eggs and milk.

   ☐   ☐   ☐

5. Mick and the Juggernauts are famous and fairly rich. Does this matter in your decision about the group paying $400 to McDonald?

   a. No, because the group is either innocent or guilty.
   b. Yes, because when innocence or guilt is unclear, then you should consider who can afford the loss.
   c. No (for another reason).
   d. Yes (for another reason).

   ☐   ☐   ☐

6. All but one of the following are reasons that McDonald wants Mick and the Juggernauts to leave.

   a. The cows produce 10 percent less milk.
   b. The chickens produce 10 percent fewer eggs.
   c. He has *never* liked rock and roll music.
   d. He doesn't sleep well because of the music.
   e. Girls walk and sleep on his land.

   ☐   ☐   ☐

7. If the milk and egg production continue to be less than normal, then

   a. the group should pay McDonald $400 a month extra.
   b. McDonald should suffer the loss because he made a mistake.
   c. McDonald should build a wall and make the group's house *soundproof.**
   d. the group should build a wall and make their house soundproof.
   e. everyone should have a big chicken and steak dinner.

   ☐   ☐   ☐

8. Bianca's parents want $1,000 for her hospital bills. They are poor people with no insurance. Who should pay?

   a. Only McDonald because he should keep his land safe.
   b. Mick and the Juggernauts because they attracted her there.
   c. Both McDonald and Mick and the Juggernauts.
   d. Bianca's parents.
   e. McDonald, Mick and the Juggernauts, and Bianca's parents.
   f. Bianca's parents and McDonald.
   g. Bianca's parents and Mick and the Juggernauts.

   ☐   ☐   ☐

built in such a way that sound is kept inside.

9. Should Mick and the Juggernauts be forced to leave?

    a. Yes.
    b. No.

□   □   □

Total Score _____

E. PRE-TRIAL QUESTIONS

Write three questions Mick is likely to ask McDonald at the trial.

1. _____
_____

2. _____
_____

3. _____
_____

Write three questions McDonald is likely to ask Mick at the trial.

1. _____
_____

2. _____
_____

3. _____
_____

F. LISTENING COMPREHENSION

Listen to the tape, then circle the correct answer.

1.   a   b   c
2.   a   b   c
3.   a   b   c
4.   a   b   c
5.   a   b   c
6.   a   b   c

G. POST-TAPE QUESTIONS

Write three questions you, as a juror, would ask Jim at the trial.

1. _____
_____

2. _____
_____

3. _____
_____

H. THE TRIAL

Defendants: Mick; McDonald.
Witnesses: Bianca; Bianca's father (or mother); Jim; McDonald's little girl.
Suits: (a.)   McDonald sues Mick for $400.
    (b.)   McDonald sues to *evict** Mick.
    (c.)   Bianca's father sues Mick and McDonald for $1,000.
Verdict: (a.)   The jury may decide on any sum, from $0 to $400. The jury chairperson may take an average of the individual jurors' decisions.
    (b.)   A vote of *yes* or *no.* Majority rules.
    (c.)   The jury must first decide the total amount to be awarded to Bianca's father. Secondly, it must decide how much each party should pay.
Procedure: Bianca and her father sit on one side of the judge. Mick, McDonald, Jim, and McDonald's daughter sit on the other side. The trial begins with the judge introducing the defendants and witnesses and reading the suits (a, b, and c) which will have been written on the board for greater clarity.

Result: (a.) _____
_____

    (b.) _____
_____

    (c.) _____
_____

"No," laughed Mr. McDonald, "a little music never hurt anyone. . . ."

*to make him move out of his house.

THE CASE OF LAZARUS

(1)   In the fourteenth century, plagues were common in Europe. It is said that one-third of the population died from the plagues. It is during a plague year that our story takes place.

(2)   Lazarus was a baker. He was a fat, lazy, and greedy man. He was married to a woman much younger who was pretty, very tall, and also a little fat. Her name was Amy, and everybody called her Big Amy. Big Amy and Lazarus had a fight one night and Lazarus left the house in a fury. He wasn't seen again for eleven months. Everyone assumed he had died of the plague. Someone said he had seen Lazarus in another city, dying of the plague.

(3)   Big Amy was in love with a lawyer named Bartolus. He, however, was not in love with her. Amy did not come from a wealthy family, but when the plague killed all of her cousins, she was the only living relative. Therefore, she inherited all of her relatives' money.

(4)   In those days marriages weren't often made for love. Bartolus saw his chance for a good marriage, but he wasn't sure that Lazarus was dead. Amy got him drunk one night and convinced him to go to the church and get married. Months later, Amy was going to have Bartolus's child.

(5)   Meanwhile, Lazarus had not died, and he had heard that Amy had become rich. He was very surprised and not at all happy to find her married to Bartolus when he returned. Amy would not let Lazarus into the house, and Lazarus said he was going to kill Bartolus.

(6)   Then the governor, who was also the judge in those days, stepped in to prevent any trouble. He called Lazarus, Bartolus, Amy, and the priest—Father Avarus —in for a trial to decide their future.

A. VOCABULARY

Circle the letter next to the answer that is similar to the word(s) on the left.

1. fourteenth century (para. 1)
   a. 1500–1600
   b. 1300–1399
   c. 1301–1399
   d. 1400
   e. 1400–1500

2. plague (para. 1)
   a. steak
   b. dog
   c. a disease, passed from one person to another
   d. a problem, passed from one person to another
   e. student

3. greedy (para. 2)
   a. fat
   b. young
   c. lazy
   d. wanting everything
   e. single

4. fury (para. 2)
   a. car
   b. joy
   c. truck
   d. anger
   e. clean suit of clothes

5. assumed (para. 2)
   a. supposed
   b. killed
   c. shot
   d. added
   e. forgot

6. wealthy (para. 3)
   a. green
   b. plastic
   c. supposed
   d. rich
   e. metal

7. relative (para. 3)
   a. dead cousin
   b. dead sister
   c. blood relation—cousin, aunt, etc.
   d. rich person who never gives money

8. marriage (para. 4)
   a. something that looks like something else
   b. sandwich
   c. coach
   d. being married
   e. soup

9. drunk (para. 4)
   a. condition of having too much beer, wine, etc. in your body
   b. sad
   c. very sick
   d. condition of having eaten too much
   e. tall

10. priest (para. 6)
    a. person who leads ceremonies in a church
    b. baker
    c. vegetable seller
    d. wine seller
    e. wine taster

Amy got drunk one night and
convinced him to go to church and get married

B. CUMULATIVE VOCABULARY EXERCISE

Fill in with words from this case, the previous cases, and the legal lexicon (Appendix 1).

When Lazarus finally came back, he and Amy were just as f _ _ _ _ _ u _ at each other as when he left.

"Why did you marry Bartolus?" he asked.

"I thought you were dead," she said.

"That was a false a _ _ _ _ _ p _ _ _ n ," he said.

"And why did you come back?" she asked. "I'll tell you why—you want my money now that I am w _ _ _ _ h _ . Your heart is full of g _ e _ _ _ . I want a divorce."

"That's impossible. Marriage is a c _ _ _ r _ _ _ which cannot be broken," said Lazarus.

"You're so old fashioned. You have a thirteenth-c _ _ _ u _ _ mind," said Amy.

C. GENERAL EXERCISES

Below is the bigamy law of the land in which Amy, Bartolus, and Lazarus live. Refer to it whenever necessary.

BIGAMY LAW

1. Every person who has a husband or wife and marries another person is guilty of bigamy.
2. This does not *include** any person
   a. whose first husband or wife is dead.
   b. whose husband or wife has been absent for one year and is assumed to be dead.
3. Every unmarried person who knows that somebody is already married but marries him/her anyway, is guilty of bigamy.

Answer the following questions according to the law. Circle your answer.

1. A single woman marries a man, knowing that he is already married. Are they both guilty of bigamy?
   a. Yes.
   b. No.

*refer to, mean.

16

2. For question 1, which part(s) of the law apply(ies)?
   a. 1
   b. 2a
   c. 2b
   d. 3
   e. 1 and 3
3. Bill's wife dies. Bill gets married later that same day. Is he guilty of bigamy?
   a. Yes.
   b. No.
4. Which part(s) of the law apply(ies) most directly to the answer for question 3?
   a. 1
   b. 2a
   c. 2b
   d. 3
   e. 1 and 3

D. LAW QUESTIONS

Answer the questions. Put the letter that corresponds to the best answer in the box marked *Your decision*. Questions 1–5 are based on the bigamy law given in section C.

1. Amy is
   a. guilty of bigamy from part 1.
   b. guilty of bigamy from part 2a.
   c. guilty of bigamy from part 3.
   d. not guilty of bigamy because a person in love is never guilty of anything.
   e. not guilty of bigamy because she is going to have a child.

Your decision   *Jury's decision*   *Score*
   ☐   ☐   ☐

2. The following (is) (are) guilty of bigamy:
   a. Lazarus and Bartolus
   b. only Bartolus
   c. Bartolus and Amy
   d. only Amy

   ☐   ☐   ☐

3. Bartolus is
   a. not guilty of bigamy because he is a lawyer.
   b. not guilty of bigamy because when you are drunk you have no responsibility.
   c. not guilty of bigamy because Amy convinced him to get married.
   d. not guilty of bigamy because it was completely the fault of the priest and Amy.
   e. guilty of bigamy.

   ☐   ☐   ☐

4. Lazarus is

   a. innocent of bigamy, but guilty of *desertion.**
   b. guilty of bigamy from part 1.
   c. guilty of bigamy from part 3.
   d. guilty of bigamy because he left Amy alone without a husband.
   e. guilty of bigamy because he was only interested in Amy's inheritance.

☐ ☐ ☐

5. The priest is

   a. guilty of bigamy from part 1.
   b. guilty of bigamy from part 2.
   c. not guilty of any crime.
   d. not guilty of bigamy, but guilty of performing an illegal marriage.
   e. guilty only in the eyes of God.

☐ ☐ ☐

6. Because of his actions, the priest should

   a. be put in jail.
   b. be sent to work in another city and not allowed to come back (banishment).
   c. be given a fine.
   d. lose his position as a priest.
   e. have to say many prayers, tell the people what he did, and ask them to forgive him.

☐ ☐ ☐

7. Amy did not get a divorce from Lazarus because

   a. Lazarus said many times he didn't want a divorce.
   b. she loved Lazarus very much.
   c. she loved Bartolus very much.
   d. she could not commit a crime.
   e. the church did not allow divorces then.

☐ ☐ ☐

8. Amy should

   a. go to jail.
   b. not go to jail because she is going to have a baby.
   c. go to jail when the child is five years old.
   d. go away and never be allowed to come back to this city (banishment).
   e. stay married to both men.

☐ ☐ ☐

9. The court should

   a. have mercy on Bartolus.
   b. have mercy on Amy.
   c. have mercy on Bartolus and Amy.
   d. have mercy on the priest.
   e. never have mercy.

☐ ☐ ☐

10. Amy says that Lazarus *deserted** her. She says that this was a serious crime. This is

   a. true, and therefore Lazarus should be put in jail.
   b. true, and therefore Lazarus should be fined.
   c. true, but Lazarus should not be jailed or fined.
   d. not true because leaving one's wife for a while is not a crime.
   e. not true because Lazarus was going to come back some time, and he has a right to leave whenever he wants and for however long he wants.

☐ ☐ ☐

11. A man has been gone for three years and has written one letter a year to his wife, but she marries again. According to the bigamy law (see section C) she is

   a. guilty of bigamy.
   b. guilty of bigamy if her first husband has been away on business or on vacation.
   c. guilty of bigamy if his letters say he is coming back.
   d. not guilty of bigamy if she thinks he is not coming back soon.
   e. not guilty of bigamy.

☐ ☐ ☐

Total Score _____

Lazarus had heard that Amy had become rich.

*leaving someone you are legally supposed to support.

*left, abandoned.

E. PRE-TRIAL QUESTIONS

You are the judge. Write two questions you would ask
Amy, Bartolus, and Lazarus.

(to Amy)

1. _____

2. _____

(to Bartolus)

1. _____

2. _____

(to Lazarus)

1. _____

2. _____

F. LISTENING COMPREHENSION

Listen to the tape, then circle the correct answer.

1.   a   b   c
2.   a   b   c
3.   a   b   c
4.   a   b   c
5.   a   b   c

G. POST-TAPE QUESTIONS

Write three questions you, as the judge, would ask the
priest (Father Avarus).

1. _____

2. _____

3. _____

Amy's inheritance is equivalent to $10,000. Who should
get the money now that Lazarus is back? How much
should each person get? Decide this as a jury (the same
group you answered the law questions with in section
D) and write the jury's decision below. Do *not* write
your personal opinion.

Amy gets _____

Lazarus gets _____

Bartolus gets _____

H. THE TRIAL

Defendants: Amy; Bartolus.

Witnesses: Lazarus; Father Avarus.

Charge: Bigamy.

Verdict: The jury should refer to the bigamy law, which the teacher has written on the board, and then decide *innocent* or *guilty*.

Sentence: Jail, or banishment, or a fine. If the jury agrees on jail or banishment, it must decide for how long it will be. If the jury decides on a fine (perhaps part of the inheritance), it must decide how much the guilty person(s) must pay.

Procedure: Along with tackling the bigamy charge, the jury should decide three related and important issues. They are: (a) who is going to be Amy's husband from now on; (b) what will happen to the baby when it is born; and (c) how will the $10,000 inheritance be divided.

Result: (a.) _____

(b.) _____

(c.) _____

(Sentence for Amy and/or Bartolus)

_____

_____

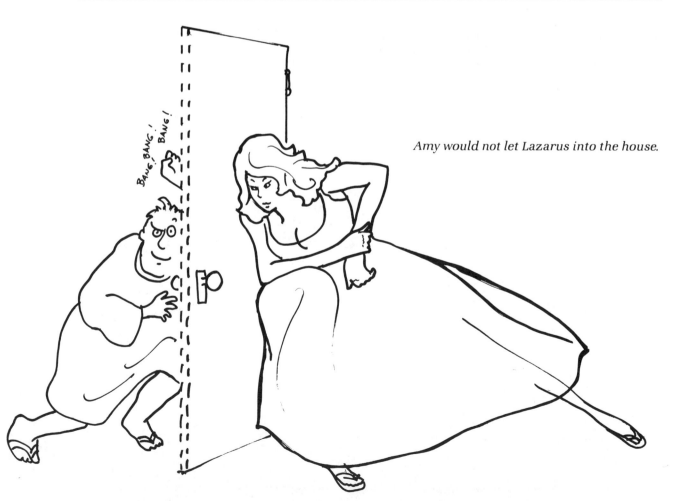

Amy would not let Lazarus into the house.

THE CASE OF FOLEY'S FOLLY

(1)   Peg Asus was a young woman who loved horses. She saved her money till she could buy a young horse named Winni. Winni became famous. She won every race she was entered in. Last year Winni was retired by Peg because she was getting too old to race.

(2)   Peg introduced Winni to a male horse named Studley who was also an excellent runner. They became fast friends and had a colt called Foley. Everyone was excited because they thought Foley was going to be a champion like his parents.

(3)   Hencher and Sparrow were young men who worked at the stables. They brushed the horses, fed them, and cleaned the stables. They saw their chance to make a lot of money. They stole the colt in the afternoon, put him in the back of their truck, and drove away. They were going to demand $100,000 for the return of the horse.

(4)   When she found out about the theft, Peg was very unhappy. She was drinking beer in a bar when she looked out the window and saw a truck go by with a horse in the back. The truck turned down the road to the stables. The road came to a dead end at the stables, just one mile ahead. There was nothing else on this road except a restaurant and a gun store.

(5)   "That's my Foley!" said Peg. She got into her Mercedes and chased after the truck, finally forcing the truck off the road. The truck hit a stone wall. Hencher and Sparrow were not hurt badly, but Foley was thrown from the back of the truck and was seriously injured. Sparrow was crying. He loved the colt. He put his head against the horse's head, and Foley bit him on the nose. Then Foley died. Sparrow was taken first to the hospital, then to jail with Hencher. Now Sparrow has an ugly scar on his nose.

A. VOCABULARY

Find words in the case that have similar meanings to those below. Write them in the blanks on the right.

1. running competition
   (para. 1)   _ _ _ _ _

2. stopped working (para. 1)   _ _ _ _ _ _ _ _

3. the best, the winner
   (para. 2)   _ _ _ _ _ _ _ _ _

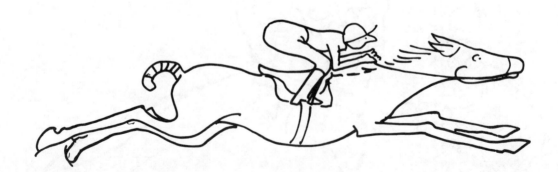

Winni won every race.

4. baby horse (para. 2)   _ _ _ _ _

5. place where horses are kept (para. 3)   _ _ _ _ _ _ _ _

6. no continuation (para. 4)   _ _ _ _ _   _ _ _ _

7. tried to catch (para. 5)   _ _ _ _ _ _ _

8. hurt (para. 5)   _ _ _ _ _ _ _ _

9. permanent mark on skin (para. 5)   _ _ _ _ _

B. CUMULATIVE VOCABULARY EXERCISE

Fill in with the singular or plural of the nouns *thief* and *theft*, or with a form of the verbs *rob* and *steal*.

Yesterday, the Barkville bank was _____ by four _____ . They _____ $50,000 from the bank. In the last month, five other banks have been _____ . The police say that the _____ are always committed by the same four _____ . One _____ has a moustache, and another has a beard. The one with the moustache also _____ handkerchiefs from anyone unlucky enough to be in the bank at the time of the _____ .

C. GENERAL EXERCISES

Match the following by drawing connecting lines.

1. Studley   a. no longer enjoys blowing his nose

2. Winni   b. father of a horse

3. Peg   c. the youngest of all

4. Sparrow   d. had a big, fast car

5. Foley   e. retired

She was drinking beer in a bar when she looked out the window and saw a truck go by with a horse in the back.

D. LAW QUESTIONS

Answer the questions. Put the letter that corresponds to the best answer in the box marked *Your decision*.

1. Who killed Foley?

   a. Hencher and Sparrow
   b. Peg
   c. Hencher, Sparrow, and Winni
   d. Peg, Hencher, and Sparrow

   Your decision   *Jury's decision*   *Score*

   ☐   ☐   ☐

2. Does Peg share part of the guilt of killing Foley?

   a. No, because she owned Foley.
   b. No, because she didn't want Foley to die.
   c. No, because she was trying to get Foley back.
   d. Yes, because she chased after the truck.
   e. Yes, because she forced the truck into the wall.

   ☐   ☐   ☐

3. Peg was in a bar before she chased the truck. This is

   a. relevant because she liked beer.
   b. relevant because she didn't like wine.
   c. not relevant until tomorrow.
   d. not relevant unless she was drunk.
   e. not relevant because she was able to see.

   ☐   ☐   ☐

Peg sues Hencher and Sparrow for $250,000 for the death of Foley. They say they shouldn't pay that amount for three reasons:
1. Winni and Studley can produce more colts.
2. Peg was probably drunk; she caused the accident.
3. There is no way to know that Foley was going to be a champion race horse. Perhaps Foley would win nothing.

---

4. Which argument is the least strong?

   a. 1
   b. 2
   c. 3

   ☐   ☐   ☐

5. Which is the strongest argument for Hencher and Sparrow?

   a. 1
   b. 2
   c. 3

   ☐   ☐   ☐

6. Hencher and Sparrow are very poor. They have no money in the bank. They were each making $10,000 a year from their jobs. They live in a cheap apartment. Suppose the court decides they should pay Peg $20,000. What should happen?

   a. Hencher and Sparrow should pay back $2,000 a year for ten years.
   b. Hencher and Sparrow should rob a bank to get the money.
   c. Hencher and Sparrow should work for two years for Peg, and she should keep half of their pay.
   d. Hencher and Sparrow have a truck, furniture, television, etc., worth $5,000. They should sell this property, give the money to Peg, and that's all.
   e. Hencher and Sparrow pay as much as they can over the next ten years, until the $20,000 is paid. If it is not all paid, after ten years they must sell all their possessions and give the money to Peg.

   ☐   ☐   ☐

7. Sparrow and Hencher should

   a. not go to jail.
   b. go to jail for six months.
   c. go to jail for one year.
   d. go to jail for three years.
   e. go to jail for five years.

   ☐   ☐   ☐

8. Foley bit Sparrow's nose. Now he looks ugly. He says it's Peg's fault. He wants $50,000 from her. Peg should

   a. pay him $50,000.
   b. pay him nothing because the accident was his fault.
   c. pay him nothing because he was already ugly.
   d. pay him nothing because he was stupid.
   e. pay him only $20,000.

   ☐   ☐   ☐

   Total Score _____

E. PRE-TRIAL QUESTIONS

Hencher and Sparrow are on trial. You are the prosecutor. Write three questions you would ask them.

1. _____

   _____

2. _____

   _____

3. _____

   _____

F. LISTENING COMPREHENSION

Listen to the tape, then circle the correct answer.

Part 1

1. a b c
2. a b c
3. a b c
4. a b c
5. a b c
6. a b c

Part 2

1. a b c
2. a b c
3. a b c
4. a b c
5. a b c
6. a b c

G. POST-TAPE QUESTIONS

You are a juror. Write two questions you would ask the bartender.

1. _____

_____

2. _____

_____

You are a juror. Write two questions you would ask Mr. Hawkes.

1. _____

_____

2. _____

_____

3. Are Hencher and Sparrow guilty of stealing the horse? (Circle your answer and discuss it with your group.)
   No, because they were bringing it back.
   No, because they were just borrowing it.
   No, because it was a baby.
   No, because it was too young to race.
   Yes.
4. Was Peg drunk? (Circle your answer and discuss it with your group.)
   Yes.
   No.
   It's impossible to know.

Hencher and Sparrow stole the colt.

H. THE TRIAL

1.	*Defendants:*	Sparrow; Hencher.
	Witness:	Mr. Hawkes.
	Charges: (a.)	Theft
	(b.)	Destruction of property (Foley).
	Verdict:	They may be innocent of one charge but guilty of another, or innocent of both, or guilty of both.
	Sentence: (a.)	If guilty, they will go to jail. The jury decides how long (chairperson takes an average).
	(b.)	If guilty, they must pay Peg some sum up to $250,000. The jury must decide how much and how Hencher and Sparrow will pay it.

Result: (a.) _____

(b.) _____

2.	*Defendant:*	Peg.
	Witness:	The bartender.
	Charges: (a.)	Drunken driving.
	(b.)	Permanent damage to Sparrow's face (disfigurement).
	Verdict: (a.)	Innocent or guilty.
	(b.)	Responsible or not responsible.
	Sentence: (a.)	If guilty, she should go to jail or pay a fine.
	(b.)	If guilty, she should pay Sparrow some sum up to $50,000: the average sum should be awarded.
	Procedure:	Both parts of the trial (1 and 2) can be done at the same time. Peg sits on one side of the judge, and Sparrow sits on the other. Since the roles of Hencher and Sparrow overlap, that of Hencher can be eliminated. If anyone really needs Hencher, he can be appointed at the last minute by the judge. Mr. Hawkes and the bartender sit next to Sparrow.

Result: (a.) _____

(b.) _____

23

THE CASE OF FARMER'S LOTS

Now he is old and can't work so hard.

(1)   Bill Farmer owned a piece of land 3,000 ft. long and 1,000 ft. wide. He divided it into three equal lots. The boundary between lots A and B is formed by a thick group of trees. The boundary between lots B and C is formed by a small stone wall.

(2)   Mr. Farmer used to farm all his land. Now he is old and can't work so hard. He farms only lot B, the lot he lives on. When he bought the land, fifty years ago, he paid very little for it. Now it is worth a lot of money.

(3)   Mr. Byers wanted to get out of the city and live on a farm. So he offered Mr. Farmer $10,000 for lot C. Farmer sold him lot C with the written agreement that the land could never be used for commercial purposes. Mr. Farmer did not like big buildings, noise, or lots of people.

(4)   One month later, Mr. Byers suddenly died, leaving his land to his son Ayer. Ayer is twenty-one and

loves living in the city. He started building a shopping center on the lot. Mr. Farmer asked the court to stop the construction.

(5)   Mr. Bilder read about this in the newspaper and went out to look at the situation. He immediately offered Mr. Farmer $100,000 for lot A. He wants to build a shopping center there.

(6)   Mr. Farmer accepted his offer. Farmer says that a shopping center on lot A would not bother him because the trees would block the noise and he couldn't see it.

(7)   Ayer is very angry. He says Farmer just wants to make a lot of money. He says he should be allowed to continue building the shopping center on lot C, before Mr. Bilder can start the other shopping center.

A. VOCABULARY

Find words in the case that have similar meanings to those below. Write them in the blanks on the right.

1. divisions of land
   (para. 1)   __ __ __ __

2. line separating
   one piece of land
   from another
   (para. 1)   __ __ __ __ __ __ __ __

3. business,
   making money
   (para. 3)   __ __ __ __ __ __ __ __ __ __

4. reasons (para. 3)   __ __ __ __ __ __ __

5. building (para. 4)   __ __ __ __ __ __ __ __ __ __ __

6. large group of
   stores (two
   words) (para. 4)   __ __ __ __ __ __ __
   __ __ __ __ __ __

7. a place, and
   how the place is
   (para. 5)   __ __ __ __ __ __ __ __ __

8. to make uncom-
   fortable, to be
   trouble (para. 6)   __ __ __ __ __ __

B. CUMULATIVE VOCABULARY EXERCISE

Fill in with words from this case, the previous cases, and the legal lexicon (Appendix 1).

Ayer was thinking about

c o __ __ __ __ __ __ __ t __ __ __ __ a t __ r __ __ __ __

for dog r __ __ __ __ s . This would attract a lot of people

and the noise would b __ __ __ __ __ __ Mr. Farmer. If he

gave away the money he made from this, it would not

be for c __ __ __ __ __ r __ __ __ __ purposes! He could

subtract this money from his taxes. He worried that his

idea might be i __ __ __ __ __ __ l . Most of all, Ayer

wanted to make enough money to r __ __ __ __ r __ at

thirty.

C. GENERAL EXERCISES

Word Exchange: Look back at paragraph 4. Write a similar but new paragraph here, filling in the blanks with different words. Your new paragraph must make sense. It may be funny if you like.

One _____ (1) later, Mr. _____ (2)
        (time)   (name)

suddenly _____ (3), _____ (4) his
         (verb)   (verb -ing)

_____ (5) _____ (6) his _____ (7)
(noun)   (prep.)   (relative)

_____ (8). _____ (9) is _____ (10), and
(name)   (same as 8)   (age)

loves _____ (11). _____ (12) started
      (verb -ing)   (pronoun for 8)

_____ (13) in _____ (14) of _____ (15). Mr.
(verb -ing)   (month)   (year)

_____ (16) asked _____ (17) to _____ (18)
(name)   (pronoun for 8)   (verb)

the _____ (19).
   (same as 13)

D. LAW QUESTIONS

Answer the questions. Put the letter that corresponds to the best answer in the box marked *Your decision*.

1. The value of Mr. Farmer's land

   a. has decreased over the last fifty years.
   b. has increased over the last fifty years.
   c. has stayed the same for fifty years.
   d. increased at first, then decreased.
   e. depends on his tomatoes.

   Your decision   *Jury's decision*   *Score*
   ☐   ☐   ☐

2. If Mr. Byers hadn't died,

   a. there probably would not be such a big problem.
   b. the value of the land would be greater.
   c. Ayer would be older.
   d. the tomatoes would be better.
   e. he would have stayed in the city.

   ☐   ☐   ☐

3. Ayer wants to build a shopping center because

   a. he loves the city.
   b. he hates the city.
   c. he likes living on a farm.
   d. he wants to make money.
   e. he hates Mr. Farmer.

   ☐   ☐   ☐

4. Mr. Farmer doesn't like

   a. Mr. Bilder.
   b. Mr. Byers.
   c. his land.
   d. the stone wall.
   e. noise and big buildings.

   ☐   ☐   ☐

25

5. Mr. Farmer sold lot C because

    a. he didn't need it for farming and he could get a lot of money for it.
    b. he didn't need it for farming and he hated neighbors.
    c. he was old and loved farming.
    d. he was old and wanted to cause problems.
    e. tomatoes grew better on lot B than on lot C.

☐ ☐ ☐

6. Mr. Farmer asked the court to stop the construction of the shopping center because

    a. he didn't like to shop.
    b. he didn't want competition for his tomatoes.
    c. he was afraid Ayer would destroy his wall.
    d. it would be noisy and ugly.
    e. he was afraid a car would hit his horse.

☐ ☐ ☐

7. Mr. Bilder thinks that a shopping center on lot A

    a. would not be very profitable.
    b. would be very profitable.
    c. would look nice.
    d. would be a good place to buy newspapers.
    e. would raise the price of tomatoes.

☐ ☐ ☐

8. Mr. Farmer is

    a. opposed to building a shopping center on lot C.
    b. opposed to building a shopping center on lot A.
    c. not opposed to building a shopping center on lot C.
    d. opposed to building a house on lot C.
    e. opposed to building anything.

☐ ☐ ☐

9. If a storm blew down all the trees between lots A and B, Mr. Farmer

    a. might move to lot C.
    b. might build a bigger stone wall between lots B and C.
    c. might divide his lot in half.
    d. might like noise.
    e. might not want a shopping center on lot A.

☐ ☐ ☐

10. Ayer is angry because

    a. Mr. Bilder might build a shopping center before he does.
    b. Mr. Bilder is a bad neighbor.
    c. city life is getting worse.
    d. his father didn't leave him enough land.
    e. lot C is not as good as lot A for farming.

☐ ☐ ☐

11. Do you think the agreement (contract) that Mr. Byers signed was a good and fair one?

    a. Yes.
    b. No.

☐ ☐ ☐

12. If Mr. Farmer dies tomorrow, Ayer should

    a. be allowed to build his shopping center.
    b. not be allowed to build his shopping center.
    c. become a farmer.
    d. make the stone wall bigger.
    e. cut down the trees between lots A and B.

☐ ☐ ☐

13. Mr. McDonald lives across the street from lots A, B, and C. When he bought his land from Farmer, twenty years ago, he signed a contract that said: "On this land you may have any animals except goats, and the land may not be used to grow tomatoes." This contract

    a. is illegal because you can put no restrictions of land use in a contract.
    b. is illegal because, although some restrictions are okay, these are ridiculous.
    c. is legal, but the restrictions end when Farmer dies.
    d. is legal, but the restrictions end if McDonald sells the property to someone else.
    e. Both c and d are correct.

☐ ☐ ☐

Total Score _____

E. PRE-TRIAL QUESTIONS

During the trial, Ayer and Mr. Farmer have an argument. Which of them makes which of the following statements? Put the name of the speaker in the blanks.

_____ 1. "Competition is good for America. You are limiting competition. It's illegal to do that."

_____ 2. "You're lying about why you stopped the construction. You just want to make money."

_____ 3. "Your father signed this contract. The contract goes with the land."

_____ 4. "The contract goes with the person who signed it. That person is dead, so the contract is no longer good."

_____ 5. "You live in the city. You don't care about country people who are disturbed by noise."

F. LISTENING COMPREHENSION

Listen to the tape, then circle the correct answer.

1. a   b   c
2. a   b   c
3. a   b   c
4. a   b   c
5. a   b   c
6. a   b   c
7. a   b   c
8. a   b   c

G. POST-TAPE QUESTIONS

Write two questions you, as a juror, would ask Mr. Bilder at the trial.

1. _____

_____

2. _____

_____

H. THE TRIAL

Defendant: Farmer.

Witnesses: Bilder; Ayer.

Suit: Ayer sues Farmer for (a) violation of his right to enjoy his property and (b) writing an illegal contract.

This is a *contracts* case. The court must decide:

(a.) If the contract goes with the land forever or if it goes only with the person who signed it.

(b.) If it is legal for a person to put a restriction in the contract. In this case the owner of lot C is restricted in his use and enjoyment of his property.

Verdict: (a.) If the jury decides that the contract goes with the land, Ayer may not build. If it goes with the person who signed it, he may build.

(b.) If the jury decides it is illegal to put any restriction of land use in a contract, Ayer may build. If restrictions are legal, he may not.

If either (a) or (b) is decided in favor of Ayer, he may build.

Procedure: Ayer sits on one side of the judge. Bilder and Farmer sit on the other. The diagram of the properties is on the board, and the players should refer to it.

Result: _____

_____

_____

_____

Ayer is 21 and loves living in the city.

THE CASE OF E.Z. RIDER

(1) Neville was a professional motorcycle racer. He loved motorcycles. One day he bought a new Honda 500cc. and rode all over town. That night he stopped at his favorite bar, Jack's, where everybody knew him. He often got so drunk there he had to call a taxi to get home.

(2) This night, Neville's friend E.Z. Rider asked for a ride home from the bar. They had an accident. Neville hit a fence on the left side of the road. E.Z. flew off the back of the motorcycle and hit a telephone pole. He broke his neck. The medical examiner said that E.Z. died instantly.

(3) Neville says a car suddenly turned into his lane, out of control. Neville also says he had to turn sharply to the left to avoid hitting the car, and then his brakes failed. The police report says there was no evidence that Neville had tried to stop.

(4) After that, Neville pushed his motorcycle two miles home. He didn't call the police to report the accident until two hours after it happened.

A. VOCABULARY

Circle the letter next to the answer that is similar to the word(s) on the left.

1. medical examiner
   (para. 2)

   a. police chief
   b. murderer
   c. undertaker
   d. person who examines bodies
   e. person who examines medicines to see if they are dangerous

2. instantly
   (para. 2)

   a. standing up
   b. once in a while
   c. from a snake bite
   d. in the water
   e. immediately

3. lane
   (para. 3)

   a. highway
   b. narrow street
   c. dead end street
   d. division in the road
   e. end of a road

One day he bought a new motorcycle

B. CUMULATIVE VOCABULARY EXERCISE

Fill in with words from this case, the previous cases, and the legal lexicon (Appendix 1).

E.Z. Rider was a c _ _ _ _ _ _ _ n motor-

cycle racer. He usually won the r _ _ _ e _ he

entered. The other drivers c h _ _ _ _ _ him but

seldom caught him. It was, however, a

p u _ _ s h _ _ _ _ sport. E.Z. was a courageous

driver. If he fell down, he would

i n _ _ _ _ _ _ _ _ get back on the bike and

continue r _ _ c _ _ _ .

C. GENERAL EXERCISES

1. *Sequencing:* Put a number (1–9) in the blank to show the order in which the events happened.

a. ____ E.Z. was found dead by the police.

b. ____ Neville phoned the police.

c. ____ Neville pushed his motorcycle home.

d. ____ E.Z. asked for a ride home.

e. ____ E.Z. hit the telephone pole.

f. ____ E.Z. died.

g. ____ Neville struck the fence.

h. ____ The medical examiner **made a report**.

i. ____ E.Z. flew off the motorcycle.

2. *Fill in with the Correct Fact:* Try to do this **from** memory.

a. The distance from the scene of the accident to Neville's house was _____ .

b. The name of the bar in which they were drinking was _____ .

c. The model of Neville's motorcycle was a _____ .

d. Instant death was the decision of the _____ _____ report.

e. Neville ran into a _____ .

f. E.Z.'s neck was broken when he hit a _____ .

Neville pushed his motorcycle 2 miles home

3. *Fill in with the Correct Preposition:*

    a. He hit a fence _____ the left side _____

       the road, and E.Z. flew _____ the back

       _____ the motorcycle.

    b. Neville rode his motorcycle all _____ town

       _____ the day _____ the accident.

    c. E.Z. asked _____ a ride _____ the bar

       _____ his house.

    d. Neville was _____ a state _____ shock

       because _____ the accident.

    e. A car turned _____ Neville's lane and Neville

       then turned sharply _____ the left.

    f. He called the police two hours _____ it

       happened.

4. *Rewrite These Sentences:* Put the words in correct order. Some may be already correct. (Do not use commas.)

    a. She was driving one night her car.

    _____

    _____

    b. She has always after she finishes work a drink at a bar.

    _____

    _____

    c. I called for my friend a taxi cab.

    _____

    _____

    d. You should never alone walk after dark in the city.

    _____

    _____

    e. The person at fault I think that is Neville.

    _____

    _____

    f. For his drinking problem Neville has himself only to blame.

    _____

    _____

D. LAW QUESTIONS

Answer the questions. Put the letter that corresponds to the best answer in the box marked *Your decision.*

1. Neville has a history of drinking problems. This is

    a. important because he was certainly drunk on the night of the accident.
    b. important because he went to Jack's.
    c. probably important, but you can't say for certain.
    d. completely irrelevant.
    e. important, unless E.Z. was drinking Coca-Cola.

Your decision	*Jury's decision*	*Score*
☐	☐	☐

2. E.Z. asked Neville for a ride home. This shows

    a. that E.Z. was certainly drunk too.
    b. that E.Z. was drunk but Neville probably wasn't.
    c. that E.Z. knew Neville wasn't drunk.
    d. only that E.Z. trusted Neville to get him home safely.
    e. that E.Z. wanted to die.

☐	☐	☐

3. Neville says his brakes failed. This is

    a. probably true because of the evidence.
    b. probably true unless Neville has a history of telling lies.
    c. probably true because he was drunk.
    d. probably not true because he hit a fence.
    e. probably not true because the motorcycle was new.

☐	☐	☐

4. If Neville is telling the truth about the car suddenly turning at him,

    a. he is innocent of killing E.Z.
    b. he is still guilty of killing E.Z.
    c. he is a liar.
    d. he is still guilty of killing E.Z. because a car is bigger than a motorcycle.
    e. he is innocent of killing E.Z. because you are never guilty when you tell the truth.

☐	☐	☐

5. There were no witnesses to the accident. Therefore,

    a. we must believe Neville.
    b. we must not believe Neville.
    c. we can never know for sure if Neville is telling the truth.
    d. we must forget the case entirely.
    e. we can never know for sure why Neville bought a motorcycle.

☐	☐	☐

6. If Neville was a very high government official instead of a motorcycle racer, we should

  a. *treat** him differently.
  b. treat him exactly the same.
  c. probably let him go free.
  d. forgive him if he promises never to drink again.
  e. take away his motorcycle.

☐   ☐   ☐

7. Neville says that E.Z. was drunk. If this is true,

  a. Neville is innocent of killing him.
  b. Neville is guilty of killing him.
  c. it doesn't matter.
  d. we must ask E.Z. his opinion.
  e. Neville is less guilty of killing him.

☐   ☐   ☐

8. Why did Neville push his motorcycle two miles home?

  a. The brakes didn't work, so he couldn't ride it.
  b. He wanted to leave no evidence that he was at the *scene of the crime.***
  c. He needed to do something to the brakes to make them not work.
  d. He was confused and in shock because of the death of his friend, and he really didn't know what he was doing.
  e. He was afraid that someone would steal it if he left it.

☐   ☐   ☐

9. Why did Neville wait two hours to report the accident?

  a. He was confused and in shock.
  b. He couldn't find the telephone number.
  c. He didn't think it was necessary.
  d. He didn't want the police to see how drunk he was.
  e. He wanted to have a good dinner before talking to anyone.

☐   ☐   ☐

Total Score _____

E. PRE-TRIAL QUESTIONS

Neville is on trial for killing E.Z. Rider. Write two questions the *prosecution* is likely to ask.

1. _____

   _____

2. _____

   _____

*act toward.   **place where the crime occurred.*

Write two questions the *defense* is likely to ask.

1. _____

   _____

2. ___ _____

   _____

F. LISTENING COMPREHENSION

Listen to the tape, then circle the correct answer.

1.   a   b   c
2.   a   b   c
3.   a   b   c
4.   a   b   c
5.   a   b   c
6.   a   b   c
7.   a   b   c

G. POST-TAPE QUESTIONS

Write one question the defense is likely to ask Neville.

1. _____

   _____

Write one question you, as a juror, would ask the bartender.

1. _____

   _____

H. THE TRIAL

Defendant: Neville.
Witnesses: Bartender; medical examiner.
Charge: Vehicular homicide (killing with a vehicle).
Verdict: If the death of E.Z. Rider was an accident, then Neville is innocent. If Neville is judged guilty, the jury must decide if there were *mitigating circumstances*— i.e., things outside of Neville's control which contributed to the death of E.Z. (For example: another reckless driver, E.Z.'s condition, etc.)
Sentence: If guilty of vehicular homicide, Neville should be sent to jail for five to ten years. If guilty with mitigating circumstances, he should be sentenced to less than five years. The jury must vote on the sentence, and the average will be the court's decision.
Procedure: Neville and the bartender sit on opposite sides of the judge. The medical examiner will sit next to the bartender.

Result: _____

   _____

   _____

   _____

   _____

WHO KILLED DENNIS DICKENS?

(1) Rick is a young reporter. He has just lost his job. He takes a long walk and then sits by a pond in a park and feeds the birds.

(2) Cher is a maid for a rich family. She is taking Dennis for a walk. Dennis is a ten-year-old boy who is badly behaved. He splashes water on Rick. Rick is angry but ignores it. Dennis thinks it's funny, so he does it again. This time Rick is really angry. It seems like the whole world is against him. He pushes Dennis into the pond and walks away. He is not worried because he knows that the pond is only one foot deep.

(3) Rick doesn't know that the caretaker, Martin Gardiner, has let holes develop in the pond. Unfortunately, at the place where Dennis falls in, the water is well over his head.

(4) Dennis cannot swim. Cher is an excellent swimmer. She can easily rescue Dennis, but she doesn't understand the danger because she is talking to her boyfriend, Deke, who is a rich, young lawyer. She does nothing but wave her hand and tell Dennis to kick his legs harder.

(5) Deke is also an excellent swimmer. He walks a few feet to the edge of the pond and reaches out to

Dennis, but Dennis is too far from the edge. Deke smiles and returns to his conversation with Cher.

(6) A few seconds later Dennis goes under and doesn't come up. He has drowned. Rick reads about the death in the *Daily News* the next day. He feels very bad. He calls the police and tells them what happened. They arrest him.

A. VOCABULARY

Circle the letter next to the answer that is similar to the word(s) on the left.

1. pond
   (para. 1)
   a. garden
   b. large body of land in which parks are found
   c. small body of water
   d. large lake
   e. swimming pool

2. splashes
   (para. 2)
   a. crashes
   b. makes
   c. prevents
   d. causes a liquid to move
   e. causes a solid to become more solid

He pushes Dennis into the pond.

3. ignores
   (para. 2)
   a. pays no attention to
   b. splashes
   c. drinks
   d. picks up
   e. puts on

4. caretaker
   (para. 3)
   a. person who washes cars
   b. person who feeds birds
   c. girl who takes care of babies
   d. person who takes care of a park
   e. person who sells carrots

5. unfortunately
   (para. 3)
   a. without luck
   b. without money
   c. like a tuna fish
   d. luckily
   e. not deep

6. rescue
   (para. 4)
   a. talk to
   b. save
   c. run away from
   d. refuse
   e. hit

7. drowned
   (para. 6)
   a. gone up and down
   b. swum
   c. floated
   d. died in water
   e. drunk water

8. *Daily News*
   (para. 6)
   a. English book
   b. comic book
   c. newspaper
   d. TV program
   e. something that happens every day

B. CUMULATIVE VOCABULARY EXERCISE

Fill in with words from this case, the previous cases, and the legal lexicon (Appendix 1).

Cher's boyfriend, Deke, used to be a lifeguard at a swimming pool. His job was to r _ _ _ c _ _ anyone who looked like he was d _ _ w _ _ _ _ _ . F _ _ _ _ _ _ _ _ _ l y , no one did.

Every Sunday there were swimming r _ c _ _ . Deke would act as the j _ _ _ _ _ of the competition. One day Rick was competing against a young man named Johannes who won the race by crossing into Rick's l_ _ _ e . Deke didn't see it and so Johannes was the winner.

Rick a c _ _ s _ _ Johannes of s t _ _ _ _ _ g the victory. Deke had to i _ _ _ _ _ e Rick's a _ _ _ _ _ _ t _ _ _ because he hadn't noticed what happened. Rick was f _ _ i _ _ _ _ , but Johannes just laughed.

"S _ _ me," said Johannes as he left with the prize.

C. GENERAL EXERCISES

1. *Fill in with* very, too, *or* enough:

a. Dennis is old _____ to walk.

b. Dennis is _____ young to swim.

c. Cher was a _____ good swimmer, but she wasn't smart _____ to realize that Dennis was in trouble.

d. Cher thought that if Dennis kicked his legs hard _____ he would be all right.

e. Dennis was _____ far from the edge for Deke to reach him.

f. Rick was _____ sad to learn about the drowning.

g. He was sad _____ to report himself to the police.

2. *Rewrite These Sentences:* Put the words in correct order.

a. Martin let develop holes.

_____

_____

b. I consider more guilty the one who pushed him.

_____

_____

c. What I think it is that was an accident.

_____

_____

d. She should never leave alone the boy.

_____

_____

e. Deke let swim the boy by himself, and returned with Cher to his converastion.

_____

_____

f. He read in the newspaper about it.

_____

_____

3. *Rewrite These Words:* See the example below.
Example: The wall was *ten feet* high.
You write: It was a ten-foot wall.

a. The snake was *ten feet long*.

_____

b. The boy was *ten years old*.

_____

c. He got a new contract *for three years*.

_____

d. The bullet left a hole in his chest *that was one inch deep*.

_____

D. LAW QUESTIONS

Answer the questions. Put the letter that corresponds to the best answer in the box marked *Your decision*.

1. Did Dennis just drown, or was he killed?

a. He just drowned.
b. He was killed.

Your decision	*Jury's decision*	*Score*
☐	☐	☐

2. Should Rick go to jail for the death of Dennis?

a. Yes.
b. No.
c. No, but he should have a different punishment.

☐	☐	☐

Deke smiles and returns to his conversation with Cher.

3. Who is *criminally responsible** for the death of Dennis?

   a. Rick only.
   b. Rick and Martin.
   c. Rick, Martin, and Cher.
   d. Rick, Martin, Cher, and Deke.
   e. Nobody.

☐   ☐   ☐

4. Rick reports himself to the police.

   a. Therefore, he is stupid.
   b. Therefore, he should go free.
   c. Therefore, he should get his old job back.
   d. This doesn't make any difference in how the court acts toward him.
   e. Therefore, the court should show mercy toward him.

☐   ☐   ☐

5. Rick reported himself to the police probably because

   a. he is stupid.
   b. he doesn't care what happens to him.
   c. he likes to have his name in the newspapers.
   d. he feels bad and doesn't want to hide anything.
   e. he thinks the court will find him innocent.
   f. he knew the police would find him anyway.

☐   ☐   ☐

6. Dennis's father, Dudley Dickens, thinks that Rick, Cher, Martin, and Deke are all responsible for the death of his child. All of the following, *except one*, are his opinions:

   a. Rick pushed Dennis into the pond.
   b. Cher didn't do her job; she was supposed to take care of Dennis.
   c. Martin didn't fix the holes in the pond.
   d. The editor of the *Daily News* unfairly took away Rick's job.
   e. Deke didn't help Dennis.

☐   ☐   ☐

Total Score _____

can be jailed as punishment.

E. PRE-TRIAL QUESTIONS

You are the judge. Rick is on trial for killing Dennis. Write two questions you would ask Rick, Cher, and Deke.

(to Rick)

1. _____

_____

2. _____

_____

(to Cher)

1. _____

_____

2. _____

_____

(to Deke)

1. _____

_____

2. _____

_____

F. LISTENING COMPREHENSION

Listen to the tape, then circle the correct answer.

Part 1
1.	a	b	c
2.	a	b	c
3.	a	b	c
4.	a	b	c
5.	a	b	c

Part 2
1.	a	b	c
2.	a	b	c
3.	a	b	c
4.	a	b	c
5.	a	b	c
6.	a	b	c

G. POST-TAPE QUESTIONS

Write one question you would ask Martin's boss.

1. _____

_____

Write one question you would ask Rick's boss.

1. _____

_____

H. THE TRIAL

The trial is divided into two parts. The second part, the suit, will take much less time because much of the pertinent testimony will already have been heard.

1. Criminal Trial

Defendant: Rick.

Witnesses: Cher; Martin; Deke; Rick's boss; Martin's boss; Mr. Dickens.

Charge: Homicide (killing of Dennis).

Verdict: In order to be found guilty, Rick must be judged directly responsible for the death of Dennis; the death must not be accidental. Even if the negligence of others contributed to the death, Rick may be judged guilty. The jury must decide if there were *mitigating circumstances* (events contributing to the death, over which Rick had no control).

Sentence: If guilty, Rick will be sentenced to 5–15 years in jail. If there are mitigating circumstances, he will receive a sentence of less than five years. The jury must vote on the sentence, taking an average.

Procedure: Rick will sit on one side of the judge. Cher, Deke, and Martin will sit on the other. Mr. Dickens, Rick's boss, and Martin's boss should also be available for questioning. The jury might have no questions at all for Mr. Dickens. As soon as the criminal action is decided, the civil suit should be undertaken.

Result: _____

_____

_____

_____

_____

2. Civil Suit

Defendants: Rick; Cher; Martin; Deke.

Witness: Mr. Dickens.

Suit: Punitive* damages of $100,000 for the death of Dennis.

Verdict: The jury must decide how much each defendant will pay. Jurors should take into consideration the ability of each defendant to pay.

Procedure: The suit is Mr. Dickens's way of punishing the people he feels were responsible for his son's death but probably would never be sent to jail for their parts in it. The jurors should question the defendants on why they should or should not pay and what their incomes are. The jury may award any sum up to $100,000; it may award less or even nothing. None of the four defendants has insurance.

Result: Amount paid by: Rick _____

Cher _____

Deke _____

Martin _____

*for punishment

He has lost his job and sits in the park and feeds the birds.

IN COLD BLOOD?

(1) Mrs. Ann Moakley lived with her two children in a large house. She got the house two years ago when she was divorced from her husband.

(2) She was going to marry her boyfriend, Bill. Bill had been living with Ann ever since the divorce. He had a terrible temper, and several times had beaten her.

(3) Although Ann loved Bill, she was afraid of him, and she persuaded him to get help from a psychiatrist, Dr. Young. One morning, at breakfast, they had an argument. Bill got up, threw his fork and knife on the table, and said to Ann, "I'll take care of you right now!" Ann threw a cup of tea at him and ran downstairs to the basement playroom where the children were watching television.

(4) A minute later, Bill opened the door at the top of the stairs and said, "If you don't come up these stairs, I'll come down and kill you and the kids." Ann started to call the police, but she hung up the phone when he started to come down the stairs. She took her rifle from the closet and held it while she started to call the police again. As soon as he came into view she fired the rifle. She was a very good shot and had practiced a lot. She put a bullet through his heart and killed him.

A. VOCABULARY

Circle the letter next to the answer that is similar to the word(s) on the left.

1. divorced
   (para. 1)

   a. taken
   b. ended a marriage
   c. forgotten
   d. given
   e. forced to live in another country because of a crime

2. beaten
   (para. 2)

   a. promised
   b. known
   c. put salt on
   d. thrown leaves at
   e. hit

3. psychiatrist
   (para. 3)

   a. doctor for bones
   b. doctor for feet
   c. plumber
   d. doctor for the mind
   e. person who is out to lunch

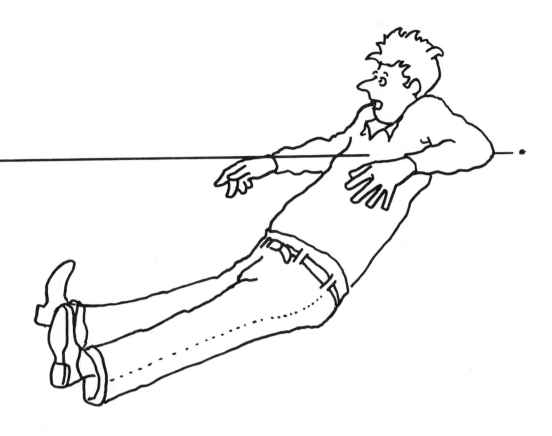

As soon as he came into view,
she fired the rifle . . .

4. take care of
   (para. 3)
   a. be nice to
   b. buy a present for
   c. make lunch and dinner for
   d. do housework for
   e. do harm to, hurt

5. basement
   (para. 3)
   a. study
   b. upstairs room
   c. room under the house
   d. place where dead bodies are taken
   e. place where only horses are kept

6. rifle
   (para. 4)
   a. kind of gun
   b. broom
   c. lunch bucket
   d. waste basket
   e. kind of ball

7. bullet
   (para. 4)
   a. piece of cake
   b. piece of metal that comes from a gun
   c. handbag
   d. can of corn
   e. horse of another color

B. CUMULATIVE VOCABULARY EXERCISE

Fill in with words from this case, the previous cases, and the legal lexicon (Appendix 1).

One night Bill heard a sound in the kitchen. He picked up his r _ _ l _ and went to see what it was. He found a man s _ _ _ l _ _ _ _ the television. Bill f i _ _ _ _ five s h _ _ _ _ at the t h _ _ _ _ who was s c _ _ _ _ _ to death. One b _ _ _ _ _ t had gone through his hat, but f o _ _ _ _ _ _ _ _ _ _ _ did not hit his head.

The man was glad when the police a _ _ _ _ _ _ _ _ d him. At the t r _ _ _ _ the d e _ _ _ _ s _ said that Bill was wild and crazy and should see a p _ _ _ _ _ i _ _ _ r _ .

C. GENERAL EXERCISES

Sequencing: Put a number (1–8) in the blank to show the order in which the events happened.

a. _____ Ann threw a cup of tea at Bill.

b. _____ Bill started downstairs.

c. _____ Ann ran downstairs.

d. _____ Bill and Ann were having breakfast.

e. _____ Ann tried to call the police.

f. _____ Ann fired the rifle.

2. *Fill in with the Correct Preposition:*

a. Ann was afraid _____ him.

b. _____ her opinion, she acted _____ self-defense.

c. Much can be said _____ defense _____ Mrs. Moakley.

d. Ann took a rifle _____ the closet.

e. She threw a cup _____ tea _____ him.

f. She shot him _____ a rifle.

g. _____ breakfast Bill threw his knife _____ the table and said he would take care _____ her.

D. LAW QUESTIONS

Answer the questions. Put the letter that corresponds to the best answer in the box marked *Your decision.*

1. Does it make any difference that Bill is not married to Ann? (Be prepared to explain your answer to your classmates.)

   a. Yes.
   b. No.

Your decision	*Jury's decision*	*Score*
☐	☐	☐

2. Dr. Young says that Bill was making progress. What does this have to say about Ann's guilt or innocence?

   a. It is completely irrelevant.
   b. It is strong evidence toward her guilt.
   c. It is strong evidence that she killed him in self-defense.
   d. It makes her seem more guilty, but it really isn't very important.
   e. It doesn't make her more guilty. Bill was not completely changed; he could still be wild.

☐	☐	☐

3. Did Ann make the right decision in not going up the stairs when Bill asked her to?

   a. Yes.
   b. No.

☐	☐	☐

4. Ann hung up the phone and didn't contact the police. Therefore,

   a. she really wasn't very afraid of him.
   b. she just made a foolish decision.
   c. she really wanted to kill Bill.
   d. she was afraid he would kill her.
   e. she knew the police would make things worse.

☐	☐	☐

5. Why didn't Ann warn Bill? Why didn't she say, for example, "Stop or I'll shoot?"

   a. She wanted to kill him.
   b. He was too close.
   c. She had no duty to warn him. Any time a person says he's going to kill you, you can shoot without warning.
   d. She was afraid he might jump at her suddenly and kill her before she could shoot.
   e. She knew he would keep coming at her, even after a warning.

☐	☐	☐

6. There is a back door from the basement to the yard. Was it a mistake not to run out with the children when Bill started down the stairs?

   a. No. Ann wanted to kill Bill.
   b. No. She has a right and a duty to defend her home.
   c. Yes. Running away would have been a safer and wiser decision.
   d. Yes. You must never shoot anybody.
   e. No. Staying and shooting was a safer way to protect herself and her children.

☐	☐	☐

7. Does it matter that Ann was a very good shot?

   a. Yes.
   b. No.

☐	☐	☐

8. Did Ann have a reason to believe that Bill had a weapon?

   a. Yes.
   b. No.

☐	☐	☐

9. Which of the following two legal expressions fits this case: "in cold blood" or "in the heat of passion"? "In cold blood" refers to something done with evil intention and without pity. "In the heat of passion" refers to something done out of extreme emotion, something done quickly and without much thought.

    a. in cold blood
    b. in the heat of passion
    c. neither
    d. both

☐   ☐   ☐

Total Score _____

E. PRE-TRIAL QUESTIONS

Ann is on trial for killing Bill. Write three questions you, as a juror, would ask her.

1. _____

_____

2. _____

_____

3. _____

_____

F. LISTENING COMPREHENSION

Listen to the tape, then circle the correct answer.

1.   a   b   c
2.   a   b   c
3.   a   b
4.   a   b
5.   a   b
6.   a   b   c
7.   a   b   c
8.   a   b   c
9.   a   b   c

Does it matter that Ann was a good shot...?

G. POST-TAPE QUESTIONS

Write two questions you would ask Dr. Young.

1. _____

_____

2. _____

_____

H. THE TRIAL

Defendant: Ann Moakley.
Witnesses: Dr. Young; one of Ann's children.
Charge: Homicide (killing of Bill).
Verdict: The jury must decide if Ann shot Bill in defense of herself and her children, or if she deliberately killed him.
Sentence: If the majority of jurors votes for self-defense, Ann is innocent and will go free. If she is found guilty of homicide, she will go to jail for ten to twenty years. The average sentence will be imposed.
Procedure: Ann sits on one side of the judge. Dr. Young and Ann's child sit on the other.

Result: _____

_____

_____

_____

_____

THE CASE OF ZANE E. GRAY

(1) Daphne Disney was a young lady of seventeen who had just finished high school. She lived in Eagan, a small town near Minneapolis, a large city in the midwest.

(2) She had always wanted to join the circus. It seemed like a very exciting life. She had promised her parents, however, that she would go to college in the fall.

(3) Daphne went to see the circus every summer when it came to Minneapolis. One day in August she was standing outside the biggest circus tent and a man dressed as a cowboy introduced himself. He was Zane E. Gray, and he said he was a knife-thrower with the circus.

(4) "There's a great trick I just learned," said Zane, "but my assistant can't be here tonight. Do you want to be my assistant for tonight? It's a little dangerous, but the pay is good."

"What do I have to do?" said Daphne.

"You stand twenty feet away with an apple on your head. I throw a knife through the apple," said Zane.

"I'm a little scared," said Daphne.

"Don't worry," said Zane, "I did it a hundred times with my old assistant."

"She's not in the hospital, is she?" said Daphne, laughing.

"No." Zane laughed too.

A man dressed as a cowboy introduced himself.

(5) He then drove her to a place in the country and successfully practiced the trick. He told her to meet him at 7:00 P.M. in front of the big tent. Daphne waited till 8:00, but Zane did not appear. She asked the circus manager where he was, and the manager said, "Zane E. Gray! He hasn't worked for us in two years, not since the terrible accident."

"What accident?" said Daphne.

"His assistant was killed. She was such a nice kid. In fact, she looked a lot like you. It was the first time Zane ever missed, and it was the last, too. I had to fire him, of course. Something like that is bad for business. The knife went right between the poor girl's eyes . . ."

(6) Daphne's eyes had closed. She fell to the ground unconscious and woke up in a hospital a day later. She stayed there for a week, taking medicine every day to calm her down. She was badly shaken by what the circus manager had said, but this was not the first time Daphne had fallen unconscious from shocking news. A year ago she had been hospitalized after serious problems with her boyfriend and her parents.

(7) Daphne's father, Mr. Disney, found Zane and asked him if what Daphne said was true. Zane laughed and said yes. Mr. Disney hit him on the nose and knocked him down. Disney said he was going to sue him for the hospital costs and for the shock to Daphne. She would miss the first semester of college.

A. VOCABULARY

Circle the letter next to the answer that is similar to the word(s) on the left.

1. tent
   (para. 3)
   a. kind of shelter made of cloth
   b. lion
   c. horse
   d. classroom
   e. library

2. cowboy
   (para. 3)
   a. student
   b. teacher
   c. horse
   d. half cow and half boy
   e. person who usually lives in the western part of the U.S. and works with cows

3. introduced
   (para. 3)
   a. killed
   b. told who somebody is
   c. hit
   d. cut
   e. taught

4. assistant
   (para. 4)
   a. donkey
   b. horse
   c. helper
   d. cousin
   e. person who is kind to small furry animals

5. unconscious
   (para. 6)
   a. sad
   b. tired
   c. dead
   d. hungry
   e. not awake

6. calm down
   (para. 6)
   a. make someone take it easy, make someone quiet
   b. make someone sick
   c. make someone do a trick
   d. make someone jump and shout
   e. make someone crazy

7. shock
   (para. 7)
   a. birthday
   b. fast loss of hair
   c. loss of desire to eat apples
   d. inability to use knives, forks, or spoons
   e. sudden violent action, like electricity

8. semester
   (para. 7)
   a. half a donut
   b. half of a football game
   c. half of a school year
   d. half an hour
   e. half an apple

B. CUMULATIVE VOCABULARY EXERCISES

Fill in with words from this case, the previous cases, and the legal lexicon (Appendix 1).

Daphne was very worried about college. She was taking difficult courses. She was afraid she would f____ "I_____ d_____ n to Chemistry" even with the a s_____ n__ of her mother who used to be a chemist.

She was beginning at the second s e_____ while all her friends had started at the first, in September. She didn't want them to know what happened. She told them she had gone camping in the w__ o___ . Her father had a l____ with a cabin on it near the b_____ a___ of Canada and the U.S. She told her friends that a wild animal attacked her and gave her a s_h_____ . She told them also that she was r__ s_c____ by a c__w____ .

It's a little dangerous, but the pay is good.

C. GENERAL EXERCISES

Word Exchange: Look back at paragraph 2. Write a similar but new paragraph here, filling in the blanks with different words. Your new paragraph must make sense. (pers. pro. = personal pronoun; ex.: he, they, etc. poss. adj. = possessive adjective; ex.: my, your, etc.)

_____ (1) had always wanted to _____ (2)
   (pers. pro.)   (verb)

the _____ (3). It seemed like a very _____ (4)
     (noun)   (adj.)

_____ (5). _____ (6) had promised
   (noun)   (same as 1)

_____ (7) _____ (8), however, that
 (poss. adj.)   (relation)

_____ (9) would _____ (10) in the
 (same as 1)   (verb)

_____ (11).
  (season)

D. LAW QUESTIONS

Answer the questions. Put the letter that corresponds to the best answer in the box marked *Your decision*.

1. Did Zane E. Gray commit a crime? (If you answer *yes*, be ready to explain your answer.)

   a. Yes.
   b. No.

   Your decision   *Jury's decision*   *Score*
   ☐   ☐   ☐

2. Is Zane E. Gray a dangerous person?

   a. Yes.
   b. No.

   ☐   ☐   ☐

3. If a person is dangerous, but did not commit a crime, she or he should

   a. be kept in jail until no longer dangerous.
   b. be kept in jail forever.
   c. be kept in a *mental** hospital until no longer dangerous.
   d. not be allowed to go out at night.
   e. be free, but be closely watched at all times by the police.

   ☐   ☐   ☐

4. If you were Daphne's mother or father, whom would you be more angry at?

   a. Daphne
   b. Zane E. Gray

   ☐   ☐   ☐

5. Mr. Disney wants Zane E. Gray to be put in jail. Zane's best defense is:

   a. He is crazy, therefore he is not responsible for anything he does.
   b. He has bad eyes.
   c. He was just having fun. None of the other girls he did the trick with went to the hospital. Daphne is the one with mental problems and he is not responsible for that.
   d. He would miss a semester at college.
   e. He could not practice his trick in jail. When he gets out he will be even more dangerous.

   ☐   ☐   ☐

6. Mr. Disney knocked Zane down by hitting him in the face. Disney should

   a. go to jail.
   b. pay Zane some money for damages.
   c. go free because he had a right to hit Zane.
   d. be shot.
   e. be Zane's new assistant for one day.

   ☐   ☐   ☐

7. Mr. Disney sues Zane E. Gray for $10,000. Should Zane pay this amount?

   a. Yes.
   b. No, he should pay nothing.
   c. No, he should pay around $5,000.
   d. No, he should pay around $1,000.

   ☐   ☐   ☐

8. What should happen to Zane E. Gray?

   a. He should go to jail.
   b. He should go to a hospital for mental patients.
   c. He should go free.
   d. He should go free if he promises to see a psychiatrist every week for a year.
   e. He should be the assistant for one day with an apple on his head while the girls whom he tricked throw the knives.

   ☐   ☐   ☐

9. Which of the following constitute(s) a criminal attack?

   a. hurting a sensitive person by using very bad (dirty) language
   b. throwing an egg at the president of the country
   c. throwing a tomato at someone one doesn't like
   d. all of the above
   e. just *a* and *b*
   f. none of the above

   ☐   ☐   ☐

*concerning the mind.

Total Score _____

E. PRE-TRIAL QUESTIONS

You are the judge. Write three questions you would ask Daphne, and three questions you would ask Zane E. Gray.

(to Daphne)

1. _____

   _____

2. _____

   _____

3. _____

   _____

(to Zane E. Gray)

1. _____

   _____

2. _____

   _____

3. _____

   _____

F. LISTENING COMPREHENSION

Listen to the tape, then circle the correct answer.

Part 1
1. a   b   c
2. a   b   c
3. a   b   c
4. a   b   c
5. a   b   c

Part 2
1. a   b   c
2. a   b   c
3. a   b
4. a   b   c
5. a   b   c
6. a   b   c

G. POST-TAPE QUESTIONS

Write one question you would ask the circus manager.

1. _____

   _____

Write one question you would ask Mrs. Disney.

1. _____

   _____

Zane admits that he has done the knife trick with a dozen girls. (Circle the answer of your choice. Discuss your answer with your group.)

1. This makes it more important that Zane go to prison or to a mental hospital.
2. This shows that Zane is really not to blame, Daphne is the crazy one, not he.

Why didn't the other girls go to the police and tell them what happened? Write your answer and discuss with your group.

_____

_____

_____

H. THE TRIAL

The trial is in two parts: criminal trial and civil suit.
1. *Criminal Trial*

Defendant:	Zane E. Gray.
Witnesses:	Daphne; Mrs. Disney; Mr. Disney; the circus manager; Daphne's old boyfriend.
Charge:	*Assault.**
Verdict:	The jury must decide two questions: (a) was Zane's conduct toward Daphne a kind of attack; (b) were Daphne's injuries the result of Zane's actions?
Sentence:	If guilty, Zane can go to jail for one to twelve months. The jury can choose an alternative sentence; for instance, a stay at a mental hospital for one to twelve months or longer.
Procedure:	Zane sits on one side of the judge. Daphne, the circus manager, Mrs. Disney, Mr. Disney, and the old boyfriend sit on the other. As soon as the criminal action is decided, the civil suit should be undertaken. The suit should be decided in about ten minutes.
Result:	_____
	_____
	_____

2. *Civil Suit*

Defendant:	Zane E. Gray.
Witnesses:	Mr. Disney; Mrs. Disney; Daphne.
Suit:	Mrs. Disney sues Zane for $10,000. Half of that is for hospital costs and half is for punitive damages for the shock and hurt to Daphne.
Verdict:	The jury may award any sum up to $10,000.
Procedure:	Each of the Disneys should say why they want $10,000. Zane should contradict them and say what he thinks is fair.
Result:	_____
	_____
	_____

*personal attack.

47

THE CASE OF CALDO

(1) Caldo was a stevedore at a big shipyard in New Jersey. He worked long hard days unloading merchandise from the big ships. He had a wife, Maria, who had come with him from Italy two years ago. She spent most of her time taking care of their two-year-old daughter, Ruth, and she had never learned much English.

(2) On August 3, 1983, after a sixteen-hour day, Caldo stopped at a bar for a few drinks. When he got home at 11:00 P.M. he found his daughter seriously ill. Maria had called Dr. Payne several times, but Mrs. Payne said that the doctor was playing golf. The fourth time Maria called, Mrs. Payne told her to stop bothering her. Maria didn't know what to do. Every time that Ruth had been ill, she had taken the child to Dr. Payne. Maria had hoped that Caldo would come home early and take care of the problem.

(3) Caldo was drunk and furious at Maria. When Caldo shouted at her, she just cried. Caldo hit her across the face and poured himself another drink. He then phoned Dr. Payne who was at home celebrating his wedding anniversary. He didn't want to see Ruth. He was tired and drunk and he wanted to go to bed. He told Caldo to take the baby to the hospital, but Caldo trusted Dr. Payne and he didn't trust hospitals. Caldo told him he was coming over with Ruth and he would break the doctor's door down if necessary. Dr. Payne was very annoyed, but he agreed.

(4) Dr. Payne said Ruth had scarlet fever, and he gave Caldo a prescription for some medicine. He made Caldo repeat the instructions on how much medicine to give and when to give it because he knew Caldo couldn't read. He also warned Caldo that too much medicine could be dangerous.

(5) Caldo bought the medicine at a drugstore, went home, and gave some to Ruth, but she vomited it up. A few minutes later he tried again, but Ruth vomited the medicine up again. He then gave her four times the correct amount to be sure that some of it would get into her blood even if she vomited again, but she didn't. An hour later Maria realized that Ruth was dead. She told Caldo so, but he was so shocked and sad that he refused to see reality. He told Maria that the medicine was not doing any good and that he was going to Dr. Payne's to get some better medicine.

(6) Dr. Payne was awakened by a loud knocking at the door. He told Caldo to go home, but Caldo said he would kill the doctor if he didn't give him some better medicine. Dr. Payne could see that Caldo was not in his right mind, so he decided to give him something that would satisfy him and would not hurt the baby. He gave him some aspirin. He was relieved to see Caldo run out because he was truly afraid that Caldo might kill him.

(7) When Caldo got home, he tried to give Ruth the new medicine and saw the letter *A* on the pills. They looked like aspirin. He took one, and as he had thought, they were aspirin. Insane with anger at the doctor, and thinking that the original medicine was also no good, Caldo took his gun and ran out the door.

(8) Maria called the doctor to warn him, but Mrs. Payne answered the phone, recognized Maria's voice, and hung up. Maria called the police, but it was too late. Caldo broke down the doctor's door and fired six shots into him, killing him.

Caldo hit her across the face
and poured himself another drink.

A. VOCABULARY

Circle the letter next to the answer that is similar to the word(s) on the left.

1. stevedore
   (para. 1)
   a. person named Steve
   b. person who digs gold
   c. a big person
   d. person who puts things on and takes things off ships
   e. a cool, dry place

2. merchandise
   (para. 1)
   a. materials, goods for sale
   b. person who sells things
   c. a cool, dry place
   d. problem
   e. medicine for cats

3. wedding anniversary
   (para. 3)
   a. the date on which someone got married
   b. the date on which someone was born
   c. the date on which someone was put in jail
   d. the year that someone went swimming
   e. pay day

4. annoyed
   (para. 3)
   a. happy
   b. hungry
   c. fat
   d. thin
   e. bothered

5. scarlet fever
   (para. 4)
   a. simple stomachache
   b. broken leg
   c. plague
   d. disease caused by old age
   e. children's sickness which causes a high temperature and redness of the skin

6. prescription
   (para. 4)
   a. something written on wood
   b. something written before lunch
   c. written permission to buy a certain medicine
   d. forbidding
   e. suitcase

7. warned
   (para. 4)
   a. stopped
   b. gave notice of danger
   c. started
   d. made hot
   e. started a fight

8. vomited
   (para. 5)
   a. made something come out of the ears or nose
   b. made something come up from the stomach
   c. made the toes fall off
   d. ate
   e. drank

9. satisfy
   (para. 6)
   a. sit for a long time
   b. injure
   c. declare
   d. make someone feel content
   e. make someone feel like he or she ate too much

B. CUMULATIVE VOCABULARY EXERCISE

Fill in with words from this case, the previous cases, and the legal lexicon (Appendix 1).

When Dr. Payne came home he was very tired. He u n _ _ _ d _ _ the car. He put his medical bag in the kitchen and his golf clubs in the b _ _ _ _ m _ _ _ . Then he parked the car in the big garage that used to be a s _ _ _ l _ for horses many years ago.

"You look tired," said his wife. "Sit down. Take a l _ _ d off your feet. I want to w _ _ n you that this crazy lady might call again. I couldn't understand what she was saying."

"What else is new?" said the doctor.

"Tomorrow I have to go to c _ _ _ _ t . Do you remember the s h _ _ _ _ i _ _ _ I saw? They want me to be a w i _ _ _ _ _ _ for the p r _ _ _ _ _ _ _ _ _ _ _ ."

Just then there was a loud knocking. It sounded like someone was b _ _ t _ _ _ the door down.

C. GENERAL EXERCISES

1. *Fill in with the Correct Fact:* (Try not to look back at the text.)

   1. Dr. Payne is having his _____ anniversary.

   2. Maria's nationality is _____ .

   3. Caldo fired _____ shots into the doctor.

   4. Ruth is _____ years old.

   5. Caldo has been in the U.S. for _____ years.

   6. Caldo works in a _____ in the state of _____ .

49

7. On the day of the killing, Caldo worked _____ hours.

8. On the day of the killing, Caldo got home at _____ .

9. The month in which the events happened was _____ .

2. *Sequencing:* Put a number (1–9) in the blank to show the order in which the events happened.

a. ____ Ruth got sick.

b. ____ Caldo shot Dr. Payne.

c. ____ Dr. Payne gave Caldo a prescription.

d. ____ Maria called to warn Dr. Payne.

e. ____ Dr. Payne said the illness was scarlet fever.

f. ____ Ruth died.

g. ____ Dr. Payne gave Caldo aspirin.

h. ____ Ruth vomited up the medicine.

i. ____ Caldo saw the letter *A* on the pills.

3. *Fill in with the Correct Preposition:*

1. Caldo was put _____ jail _____ killing Dr. Payne.

2. Caldo has lived _____ the U.S. _____ two years.

3. Caldo worked _____ a stevedore _____ a big shipyard _____ New Jersey.

4. Dr. Payne stopped _____ a bar _____ work.

5. Dr. Payne gave him a prescription _____ medicine.

6. Mrs. Payne was sick _____ talking _____ Maria.

7. Mrs. Payne hung _____ the phone _____ soon _____ she heard Maria's voice.

8. Caldo was insane _____ anger _____ the doctor.

4. *Answer True or False (T or F):* Read the case once more, but do not refer to it while answering.

1. ____ Caldo worked sixteen hours on the day of the killing.

2. ____ Dr. Payne spoke very little English.

3. ____ Caldo went directly home after work.

4. ____ Dr. Payne made Caldo repeat the instructions for the medicine because Caldo had a bad memory.

5. ____ Caldo knew that too much of the medicine could be dangerous.

6. ____ Ruth vomited after Caldo gave her four times the amount of medicine he was supposed to give her.

7. ____ Caldo realized immediately that Ruth was dead.

8. ____ Maria actually warned Dr. Payne.

9. ____ Caldo shot the doctor with one bullet.

D. LAW QUESTIONS

Answer the questions. Put the letter that corresponds to the best answer in the box marked *Your decision.*

1. Who is responsible for Ruth's death?

   a. Dr. Payne
   b. Dr. Payne and Caldo
   c. Dr. Payne, Caldo, and Maria
   d. Maria and Caldo
   e. Caldo

   Your decision   *Jury's decision*   *Score*
   ☐   ☐   ☐

2. Who should go to jail for the death of Ruth?

   a. no one
   b. Caldo
   c. Maria and Caldo
   d. Maria
   e. Mrs. Payne

   ☐   ☐   ☐

3. If Dr. Payne had not died, he should

   a. be shot again.
   b. lose his license to practice medicine.
   c. go to jail.
   d. pay a fine.
   e. be warned that he must be more careful.

   ☐   ☐   ☐

50

4. The defense says that Caldo was *temporarily insane** and therefore was not responsible for his actions. Do you agree?

a. Yes.
b. No, he knew perfectly well what he was doing.
c. Yes, he was temporarily insane, but he was still responsible for his actions.

☐ ☐ ☐

5. Is Maria also *criminally responsible*** for Dr. Payne's death?

a. Yes, because she speaks bad English.
b. No, because she speaks bad English.
c. Yes, because she was able to warn Dr. Payne but she didn't.
d. No. Not warning someone is not a crime.
e. No, because she was temporarily insane.

☐ ☐ ☐

6. If Ruth didn't have scarlet fever,

a. Caldo would be right to shoot Dr. Payne.
b. Caldo's crime would be less serious.
c. Caldo's crime would be more serious.
d. it wouldn't make any difference.

☐ ☐ ☐

7. The prosecution says that this is a very serious crime because Caldo *intended**** to kill the doctor. The defense says that the killing was not intentional; it was done in the heat of passion. Who is right?

a. the prosecution
b. the defense

☐ ☐ ☐

Total Score _____

E. PRE-TRIAL QUESTIONS

Write two questions you would ask Caldo, two questions you would ask Maria, and one question you would ask Mrs. Payne.

(to Caldo)

1. _____

   _____

2. _____

   _____

(to Maria)

1. _____

   _____

2. _____

   _____

(to Mrs. Payne)

1. _____

   _____

F. LISTENING COMPREHENSION

Listen to the tape, then circle the correct answer.

1. a b b   5. a b b
2. a b c   6. a b c
3. a b c   7. a b c
4. a b b

G. POST-TAPE QUESTIONS

Write two questions you would ask the doctor who examined the body of Ruth.

1. _____

   _____

2. _____

   _____

H. THE TRIAL

Defendant: Caldo.
Witnesses: Mrs. Payne; Maria; the medical examiner.
Charge: Murder.
Verdict: If Caldo killed Dr. Payne in cold blood (deliberately and intentionally), he is guilty of murder. If there are mitigating circumstances, the jury may decide to vote him guilty of homicide—a lesser crime. Some jurors may feel that he is innocent because killing Dr. Payne was justified.
Sentence: If guilty of murder, the sentence will be death or life imprisonment. If guilty of homicide, the sentence will be 5–20 years in jail.
Procedure: Caldo sits on one side of the judge, and Mrs. Payne, Maria, and the medical examiner sit on the other.

Result: _____

   _____

   _____

   _____

   _____

*crazy, out of his mind, for a short period of time.
**should be put in jail for.
***planned.

51

When she was nervous or felt bad, Karen ate sweet things

THE CASE OF KAREN HENDERSON

(1) Karen Henderson was a French-Canadian who came to the U.S. thirty years ago. She was fifty-eight years old and had worked as a secretary to the president of the Mathers Candy Company for twenty-four years. Karen made only $5,000 a year when she started, but she got a $1,000 raise every year, until she she was making $29,000. On May 1, 1983, the president of the company, Mr. Mathers, told her she was no longer needed. He was going to replace her with a new secretary for $10,000. Karen was very upset because she was going to be sixty years old in two years and could retire from the company with a pension of $5,000 a year. Now she would get nothing.

"Is this the way you repay me for twenty-four years of service?" Karen asked.

"You have one more month; then you're finished," said Mr. Mathers. "You don't work as fast as you used to, and besides, you've missed ten days due to illness in the last five months. I need a fast and healthy secretary. I'm sorry, Karen, but business is business. Sooner or later age must step aside and let youth take over. None of us is getting any younger."

(2) Karen saw her lawyer and they decided to sue the company. It seemed so unfair. They were going to argue that the company was practicing illegal age discrimination. Karen and her lawyer felt that it was illegal to fire someone just for getting old.

(3) Karen had health problems, and this bad news made them worse. She had diabetes, a disease of the blood. She had to take insulin every day. She was not supposed to eat sweet things, but when she was nervous and felt bad, she ate them anyway. She could not stop herself. Once, after a fight with her husband, Jack, she ate a pound of chocolates and had to to go the hospital.

(4) When the company found out that Karen was going to sue, Mr. Mathers told her she could stay. He gave her a new office—it used to be a closet! It had no windows and was very small. She was there by herself, and Mathers took away all of her responsibilites. Now she had almost nothing to do all day. Once, when she was crying, Mr. Mathers brought her a pound of chocolates. She ate them all. She got sick and almost died. Mathers told the other employees that he didn't know she was a diabetic, and he said that he didn't know that diabetics weren't supposed to eat sweets.

(5) Karen knew that if she quit she would get no pension and she would not be able to sue. She was being forced to leave, but she wanted to get her revenge before she left. She knew that ten years ago the company was in deep financial trouble. In 1972 Mr. Mathers changed the company's books so that they paid no taxes. The company was supposed to pay $25,000. Karen told Mathers that she was going to tell the truth to the newspapers and to the police. She did not like doing that, and she left Mathers's office crying.

(6) At the same time, she was having problems with her marriage. Her husband, Jack, also worked for the Mathers Candy Company, and he was interested in Mathers's new secretary, Sadie. Sadie had already taken over many of Karen's old responsibilities.

(7) Karen decided to delay talking to the newspapers. Four days later, on May 10, 1983, her husband found her dead in bed. The medical examiner said she had died from too much sugar in her blood. By her bed was an empty box of chocolates and an empty half-gallon container of chocolate ice cream.

(8) The police found a typewritten suicide note. They discovered that two pages of her diary had been torn out. One of the pages was from May 7, 1983, and the other was from May 3, 1972. Parts of the diary were written in French. These parts told the truth about Mathers not paying taxes.

A. VOCABULARY

Circle the letter next to the answer that is similar to the word(s) on the left.

1. retire
   (para. 1)
   a. put new tires on a car
   b. sleep
   c. pull back
   d. stop working after years of service
   e. stop smoking

2. pension
   (para. 1)
   a. hotel
   b. apartment
   c. amount of money you get when you stop working
   d. a watch
   e. amount of money you win playing cards

3. discrimination
   (para. 2)
   a. hatred of children
   b. killing of old people
   c. choosing one person or thing over another
   d. treating everybody the same
   e. a country where disco music is loved

4. insulin
   (para. 3)
   a. medicine taken for diabetes
   b. material protecting against heat
   c. a sweet
   d. coffee break
   e. good news

5. diabetics
   (para. 4)
   a. a disease
   b. terrible
   c. secretaries
   d. people who have diabetes
   e. people who don't have diabetes

6. quit
   (para. 5)
   a. stop eating
   b. stop taking medicine
   c. start eating
   d. stop working
   e. silent

7. revenge
   (para. 5)
   a. chocolates
   b. doing something to somebody who has done you wrong
   c. buying a present for someone you like
   d. killing
   e. disease

Her husband was interested in Mathers's new secretary

8. financial
   (para. 5)

   a. concerning fun
   b. relative
   c. concerning insects
   d. concerning money

9. books
   (para. 5)

   a. accounting records
   b. disco records
   c. children's literature
   d. paper wrapper around a
      candy bar
   e. list of addresses

10. suicide
    (para. 8)

    a. killing oneself
    b. recipe
    c. birthday
    d. happy

11. diary
    (para. 8)

    a. English book
    b. French book
    c. notebook in which you write
       what happens each day
    d. disease
    e. telephone directory

B. CUMULATIVE VOCABULARY EXERCISE

Fill in the blanks using vocabulary from this case, the previous cases, and the legal lexicon (Appendix 1).

(This is the entry from Karen's diary for April 10, 1983.)

Dear D _ _ _ _ _ _ ,

I can't wait for my r _ _ _ _ _ _ _ _ _ _ t .

Taxes are due in five days, and all I do is check the

b _ _ _ s .

I am so tired of my d _ _ _ b _ _ _ _ _ ! I

almost q _ _ _ _ taking my i _ _ _ _ _ _ _ n .

I had a c _ _ _ z _ dream last night in which

Jack was eating my ice cream. I said, "You can't have

your ice cream and eat it too." I don't know why I said

that. The ice cream had two flavors—vanilla and

chocolate, but Jack was eating only the chocolate. I

said, "You can't d _ _ _ r _ _ _ n _ _ _ like

that!" He felt g _ _ _ l _ y . I got my

r _ _ _ _ _ g _ by forcing him to eat the vanilla too.

He didn't like it—not one bit.

C. GENERAL EXERCISES

1. *True or False:*   Write *T* or *F* in the blanks.

   1. ____ Karen Henderson was president of the
            Mathers Candy Company.

   2. ____ Karen started working for Mathers
            immediately upon coming to the U.S.

3. ____ Secretaries always start at $5,000/yr.

4. ____ If Karen retires at sixty she will get a
         pension.

5. ____ If Karen quits at fifty-eight she will get a
         pension.

6. ____ The company was forcing Karen to leave.

7. ____ Mr. Mathers no longer needs a secretary.

8. ____ Karen's last office was nice.

9. ____ Karen never missed a day of work.

10. ____ Karen quit.

11. ____ Mr. Mathers became afraid of Karen.

12. ____ Karen wrote her diary entirely in English.

13. ____ Knowledge can be dangerous.

2. This is a copy of the remaining part of the page torn out of Karen's diary.

May 7, 1983
Dear Diary,
   Yesterday I had a talk with the boss.
I told him I was going to tell the new
about how the company never pa

How do you think this sentence probably finished?

   I told him I was going to t _ ll the

   new _____ about how the company never

   p _____

   _____ .

3. This is a copy of the page Karen wrote in French in 1972.

   May 3, 1972

   Cher Journal,

      J'ai un secret terrible. _____

   _____

   _____ .

Two of the words in the sentence are the same in English. Take a guess at translating the entire sentence into English.

_____

_____

How do you think the page continued? What was the secret she referred to?

_____

_____

_____

_____ .

D. LAW QUESTIONS

Answer the questions. Put the letter that corresponds to the best answer in the box marked *Your decision*.

1. Did Mr. Mathers make a good decision when he told Karen she was no longer needed?

   a. Yes, because he is a businessman and it was a smart business decision.
   b. Yes, because a young person can work better than an old person.
   c. No, because it was not true: she really was needed.
   d. No, because it was unkind and unfair.
   e. No, because he forgot that she could get revenge.

   Your decision   *Jury's decision*   *Score*
   ☐   ☐   ☐

2. Should there be a law against age discrimination?

   a. Yes, because old people need job protection.
   b. Yes, because young people need job protection.
   c. No, because age doesn't make any difference in any job.
   d. No, because the employer should be free to employ anyone she or he wants.
   e. No, because we have too many laws already.

   ☐   ☐   ☐

3. Should Mr. Mathers consider Karen's health problem in his decision to let her go?

   a. No, that's entirely her personal problem.
   b. Yes, because working hard could make her health much worse and put her life in danger.
   c. Yes, because it's important for the company that she be healthy.
   d. No, because she probably has health insurance.
   e. No, because there is no direct connection between diabetes and losing a job.

   ☐   ☐   ☐

4. Was Mr. Mathers being kind by bringing Karen chocolates?

   a. Yes, he was trying to make her feel better.
   b. Yes, because you can't cry when you are eating.
   c. Yes, it's always nice to give someone candy.
   d. No, he must know that chocolates make her sick.
   e. No, because they were poisoned.

   ☐   ☐   ☐

5. Was Karen right in trying to get revenge?

   a. Yes. As the Bible says, "An eye for an eye, a tooth for a tooth." They hurt her, so she should hurt them.
   b. No. You should never try to get revenge.
   c. Yes. It would teach the company a lesson, and they would not be so quick to get rid of old people.
   d. No, because she would also hurt a lot of good people who work for the company.
   e. Yes, because Mr. Mathers is a bad man.

   ☐   ☐   ☐

6. Who might be happy to have Karen dead?

   a. Jack Henderson
   b. Mr. Mathers
   c. both Jack Henderson and Mr. Mathers
   d. the police chief
   e. Karen's lawyer

   ☐   ☐   ☐

7. Which of the following is a likely description of the May 3, 1972, page missing from Karen's diary?

   a. a recipe for Chinese fish
   b. how Mathers changed the books—written in French
   c. how Mathers changed the books—written in English
   d. how she tried to commit suicide by eating the chocolates that Jack gave her after a fight with him
   e. both *c* and *d* are likely

   ☐   ☐   ☐

8. Which of the following is a likely description of the May 7, 1983, page missing from Karen's diary?

   a. what she and her lawyer talked about
   b. what she and her doctor talked about
   c. her retirement plans
   d. her conversation with Mathers in which she told him she was going to tell the police he didn't pay the taxes he was supposed to
   e. ideas about new jobs she might look for

   ☐   ☐   ☐

9. Which of these people had no reason to kill Karen?

   a. Mr. Mathers
   b. Jack Henderson
   c. Sadie
   d. both *b* and *c*
   e. all of them

   ☐   ☐   ☐

55

10. What do you think is the most probable cause of Karen's death?

    a. She committed suicide.
    b. Jack murdered her by giving her chocolates and ice cream.
    c. Jack and Mr. Mathers together planned to give her chocolates and ice cream in order to kill her.
    d. Sadie, Jack, and Mr. Mathers *conspired** to kill her.
    e. She died of natural causes.

☐   ☐   ☐

11. If Karen was murdered, was she partly to blame for her own death?

    a. Yes, because she was weak and ate sweets when she felt bad, knowing they could kill her.
    b. Yes, because she was "asking for trouble" in taking revenge on Mr. Mathers.
    c. both *a* and *b*
    d. No, a person is never to blame for her or his own murder, even partly.
    e. No.

☐   ☐   ☐

Total Score _____

E. PRE-TRIAL QUESTIONS

Write two questions the prosecution is likely to ask Jack.

1. _____

_____

2. _____

_____

Write two questions the prosecution is likely to ask Mr. Mathers.

1. _____

_____

2. _____

_____

F. LISTENING COMPREHENSION

Listen to the tape, then circle the correct answer.

1.	a	b	c	7.	a	b	c
2.	a	b	c	8.	a	b	c
3.	a	b	c	9.	a	b	c
4.	a	b	b	10.	a	b	c
5.	a	b	c	11.	a	b	c
6.	a	b	b				

*met secretly and planned to do something bad.

56

G. POST-TAPE QUESTIONS

Using the new facts you have learned from the tape, write two more questions the prosecution is likely to ask Jack.

1. _____

_____

2. _____

_____

Write two more questions the prosecution is likely to ask Mr. Mathers.

1. _____

_____

2. _____

_____

H. THE TRIAL

Defendants: Jack Henderson; Mr. Mathers.
Witnesses: Sadie; Captain DesMaisons.
Charge: Murder.
Verdict: The jury must decide whether both are guilty or just one of them is guilty. Of course, both may be found innocent. In order to be found guilty of murder, the defendant(s) must have deliberately killed Karen; it cannot have been an accident.
Sentence: If guilty of murder, the defendant(s) will be sentenced either to death or to life imprisonment.
Procedure: Jack and Mathers sit on one side of the judge. Sadie and Captain DesMaisons sit on the other. Remember that Captain DesMaisons can read French. He should be asked to translate what was written in the diary in French on May 3, 1972 (cf. C. *General Exercises*, question 3).

Result: _____

_____

_____

_____

_____

APPENDICES

LEGAL LEXICON

1. **to accuse** (v.)—To say that someone did something wrong. (Note: "Of" is generally used with "accuse.") Example: Two men *accused* Clyde *of* stealing money.
2. **to arrest** (v.), **arrest** (n.)—To stop someone and take him to jail. Example: The police *arrested* Clyde on Friday.
3. **to commit a crime** (v.)—To take part in a crime. Example: Clyde *committed a crime* when he stole the money.
4. **court** (n.)—The place where an accused person defends her- or himself. Example: Clyde will be taken to *court* on Monday.
5. **defendant** (n.), **to defend** (v.)—The person who is accused of a crime. Example: Clyde is a *defendant*.
6. **guilty** (adj.), **guilt** (n.)—Having committed a crime. Example: Clyde is *guilty* of stealing money.
7. **innocent** (adj.), **innocence** (n.)—Not having committed a crime. Example: Bonnie says she didn't steal the money. She says she is *innocent*.

8. **jail** (n.), **to jail** (v.)—Prison. Example: They put Clyde in *jail*.
9. **judge** (n.), **to judge** (v.)—The officer who runs the court; the head of the court. Example: The *judge* decided that Clyde should spend one year in jail.

The jury

10. **jury** (n.), **juror** (n.)—The group of twelve people who listen to a case and decide if the accused person is guilty or innocent. Each of these twelve people is a *juror*. Example: The *jury* told the judge they all agreed that Clyde was guilty.

11. **legal** (adj.), **illegal** (adj.) = opposite of legal—Within the law; lawful. Example: It is *legal* to drive your car at 55 m.p.h., but it is *illegal* to drive at 70 m.p.h.

12. **mercy** (n.)—Pity; being kind when deciding against someone. Example: The judge thought Clyde was a bad man, so he had no *mercy* on him. He gave him the maximum time in jail.

13. **prosecution** (n.), **prosecutor** (n.), **to prosecute** (v.)—Taking action against a criminal in court. Example: The *prosecution* asked the judge to put Clyde in jail for one year.

14. **to punish** (v.), **punishment** (n.)—To make someone suffer. Example: The court *punished* Clyde by putting him in jail.

15. **relevant** (adj.), **irrelevant** (adj.) = opposite of relevant—Related to something else; having to do with something. Example: Mrs. Defarge had very bad eyes. This is *relevant* to her knowing it was Clyde who robbed the bank.

16. **to restrict** (v.), **restriction** (n.)—To limit. Example: Swimming is *restricted* to just one beach because the others are too dangerous.

17. **to rob** (v.), **robbery** (n.)—To steal; to take something from someone without permission. Example: Clyde *robbed* the bank. (Note: The difference between *rob* and *steal* is that a criminal *robs* a store, a bank, a place, while we say that a criminal *steals* something—a watch, a ring, some money. "From" is often used with "steal"; example: Clyde *stole* some money *from* the bank, but Clyde *robbed* the bank. You cannot say, "Clyde *stole* the bank." This means that Clyde picked up the bank and ran away with it!)
    robber (n.)—A person who steals. Example: Clyde is a *robber*.

18. **sentence** (n.), **to sentence** (v.)—The time a person spends in jail for committing a crime. Example: The judge gave Clyde a *sentence* of five years.

19. **to sue** (v.), **suit** (n.)—To ask the court to make someone pay for damages. Example: Clyde broke Mrs. Defarge's arm, so she *sued* him for the hospital costs—$500.

20. **thief** (n.)—Robber; a person who steals. Example: Clyde is a *thief*.
    theft (n.)—The act of stealing. Example: Yesterday there was a *theft* at the Barkville bank.

21. **trial** (n.)—What happens in the court—the defendant speaking for him- or herself and the prosecution speaking against the defendant. Example: Clyde went to *trial* for robbing the bank.
    to try (v.)—To put someone on trial. Example: On Monday the prosecution will *try* Clyde for robbing the bank.

22. **verdict** (n.)—The jury's decision of guilty or innocent. Example: Judge: "What is your *verdict*?" Jury: "We find the defendant guilty."

23. **witness** (n.), **to witness** (v.)—A person who saw a crime being committed. Example: Mrs. Defarge was a *witness* against Clyde.

LISTENING COMPREHENSION TAPESCRIPTS

CASE 1

Wolgang goes to a gun store. Listen to the dialogue between him and the gun store clerk.

Clerk:	May I help you?
Wolfgang:	Yes. I want to buy a . . . a . . .
C:	A gun?
W:	Yes.
C:	You're a foreigner, aren't you?
W:	Yes. I'm from Germany. Will you please hurry up and sell me a gun?
C:	Are you angry about something?
W:	Would I buy a gun if I was happy? Of course I'm angry. And I'm afraid. This big gorilla said he was going to kill me! And he said he was going to break my hands!
C:	(laughing) Well, if you're dead it doesn't matter if your hands are broken.
W:	I teach piano. I depend on my hands to make a living.
C:	OK. Take it easy. I'm not supposed to sell a gun to anyone who appears angry. It's against the law.
W:	Very well, I'll go to another store.
C:	Hold on! Don't be in such a hurry. How much do you want to spend?
W:	All I have is $30. That's what I get every week from my student.
C:	Here's my cheapest gun—$30.
W:	Good. Here is your money. Now I won't be able to buy food until he pays me next week.
C:	I sure hope he pays you, mister. You look like you'd shoot him if he didn't. (laughs)

Questions
1. What word does Wolfgang have trouble remembering?
   a. please
   b. gun
   c. buy
2. Where is Wolfgang from?
   a. Austria
   b. Russia
   c. Germany
3. How does Wolfgang feel?
   a. sad
   b. angry and sad
   c. angry and afraid
4. It is illegal to
   a. sell guns.
   b. sell guns to foreigners.
   c. sell guns to angry people.
5. How much does Wolgang make per week?
   a. $13
   b. $30
   c. $300
6. What will happen if Ludwig doesn't pay Wolfgang next week?
   a. Wolfgang won't be able to pay for the gun.
   b. Wolfgang won't be able to buy food.
   c. Wolfgang won't be able to buy a piano.
7. Is it easy to buy a gun in America?
   a. No.
   b. Yes, but only if you are rich.
   c. Yes, but only if you speak good English.
   d. Yes.

CASE 2

Joe, who lives behind Zitkopf and Links on Route 14, is visiting his old friend Farmer Brown in the city. Listen to their conversation.

Brown:	So, what's new, Joe?
Joe:	You'll never guess who came back?
B:	Who?
J:	Links!
B:	I thought he'd moved to Canada for good.
J:	I wish he had. You remember what a troublemaker he was.
B:	Do I! He was in jail more than he was out. Always getting into fights. Stealing chickens and lettuce.
J:	Well, you know how it is when you're young.
B:	Guys like that never change, Joe. Once a troublemaker, always a troublemaker.
J:	Maybe you're right. He just drove his truck through Zitkopf's garden.
B:	Ha! Zitkopf deserves it. He knew Links had the right of way, but he built that swimming pool anyway. He's a fool.
J:	I agree with you there. He says he's going to sue Links for the 20 heads of lettuce he destroyed. What's that—$10? But I tell you, I could solve both of their problems.
B:	How is that?
J:	Links could use my driveway. It goes right near his land. Of course, I don't like anybody using my driveway, but insects ate all my corn this year. I'd sell the right of way through my land for $1,000.

Questions
1. Where was Links living before he came back?
   a. Cambodia
   b. Canada
   c. Philadelphia
2. What did people think of Links?
   a. He was a troublemaker.
   b. He was a good farmer.
   c. He was an excellent swimmer.

3. Joe and Mr. Brown both think Zitkopf is
   a. a troublemaker.
   b. a fool.
   c. an excellent swimmer.
4. What is the value of 20 heads of lettuce?
   a. $20
   b. ten cents
   c. $10
5. Joe suggests Links could get to his land by using
   a. his garden.
   b. his path.
   c. his driveway.
6. Joe will sell the right of way through his land for
   a. $20.
   b. $1,000.
   c. $2,000.
7. Why is Joe willing to sell the right of way?
   a. because he likes people using his driveway
   b. because he likes people
   c. because he needs the money
8. This year Joe's farm was
   a. successful.
   b. unsuccessful.
   c. sold.

Questions
1. Why didn't Mr. McDonald close his window?
   a. because he likes music
   b. because he likes to look at the stars
   c. because he likes fresh air
2. What does Jim think of musicians from the city?
   a. They're nice people.
   b. They're bad people.
   c. They're sour people.
3. Jim's chickens and cows are giving
   a. fewer eggs and less milk.
   b. more eggs and milk.
   c. the same amount of eggs and milk.
4. On what does McDonald blame the lower production of eggs and milk?
   a. the heat
   b. the musicians
   c. the chickens and cows
5. Jim thinks the musicians have
   a. a lot of money.
   b. little money.
   c. less money than McDonald.
6. Jim thinks that maybe Bianca broke her leg by
   a. falling off her motorcycle.
   b. falling off Mick's horse.
   c. falling off Mick's motorcycle.

CASE 3

Jim, a neighbor, comes to visit the McDonalds, who are eating breakfast.

	(knock)
McDonald:	Jim, come in.
Jim:	Hi! What's new?
M:	Couldn't sleep again last night. That music kept me awake.
J:	Why don't you close your window? Then it won't be so loud.
M:	I like fresh air. Why should I have to close my window They shouldn't be playing at that hour.
J:	I know how you feel. Those musicians are no good. They're city people. They don't belong here. And all those girls coming out here to listen to them—at all hours of the night! It's terrible!
McDonald's little girl:	Daddy! Daddy! My milk is sour!
M:	You see, Jim, that music upsets my cows so much they give sour milk.
J:	Ha! Your milk never was any good.
M:	Nonsense. It's usually good. It was good until those musicians came here. And now the cows are giving less milk and the chickens are laying fewer eggs.
J:	So are my cows and chickens. Maybe it's just the heat.
M:	I don't know what I'm going to do. I don't have any money, and now that girl Bianca is suing me.
J:	Those musicians should pay. It's their fault and they have a lot of money. Besides, I saw her fall off Mick's motorcycle. Maybe that's how she broke her leg.

CASE 4

Amy wants to marry Bartolus. Listen to this conversation between Amy and Father Avarus, a priest.

Amy:	Father Avarus, you must help me!
Avarus:	What can I do for you, my child?
Amy:	I'm in love with Bartolus and I must marry him.
Avarus:	But you are already married to Lazarus.
Amy:	He's been gone for eleven months. I'm sure he's dead. Half the people in this city have died. How could he be alive? Why wouldn't he come home if he were alive?
Avarus:	I'm sorry, but you know the law. He must be gone for one year before you can marry again. Why don't you wait one month?
Amy:	Father, you don't understand love. Listen, I have become wealthy. You are so poor. All you eat is old bread and cheese. I'll give you a lot of money if you marry us. Take it! (sound of coins) Life is so short, and you'll probably die of the plague. Enjoy life while you can.
Avarus:	Perhaps you're right. I hate to take this money, but I do need a new horse. My old horse just died of the plague. We can pay this man I know to say he saw Lazarus dead. Then I can marry you and Bartolus.
Amy:	What a great idea!
Avarus:	Still I don't understand why you're in such a hurry to get married again. You know that the husband legally owns everything except the bed! All your other property will belong to Bartolus.
Amy:	Father Avarus, you just don't understand love.

Questions
1. How many people in this city have died from the plague?
   a. 5%
   b. 50%
   c. 15%
2. Father Avarus suggests that Amy wait for
   a. one day.
   b. eleven months.
   c. one month.
3. Why does Father Avarus take the money from Amy?
   a. because he needs better bread and cheese
   b. because he needs a new horse
   c. because he needs a longer life
4. Father Avarus will pay a man to be
   a. a false witness.
   b. a wealthy man.
   c. the owner of a horse.
5. Amy thinks that Father Avarus does not understand
   a. religion.
   b. love.
   c. money.

CASE 5

Part 1
Sparrow and Hencher are at Sparrow's house. It is the morning after they stole the horse. Sparrow's father, Mr. Hawkes, is making breakfast.

Mr. Hawkes: How about some eggs and bacon, boys?
Sparrow: Eggs! Dad, we're going to have champagne!
Hencher: Yes, Mr. Hawkes, we're going to be rich!
Mr. H: What's this all about?
S: Dad, we stole Foley, the racehorse. We're going to ask Peg for $100,000 for its return.
Mr. H: That's crazy! You're the first ones the police will look for. You'll go to jail.
H: Maybe your father is right, Sparrow. As soon as we start spending a lot of money, they'll know it was us.
S: What if we go to California?
H: They'll follow us, I'm sure.
Mr. H: You boys have made a big mistake. But you still have time to fix things. You haven't asked for the money yet, so why don't you bring the horse back? It's been gone less than twenty-four hours. And if somebody sees you, you can just tell them you were borrowing it—you wanted to take a ride somewhere.
H: Yes! Sparrow, let's do it! I'm scared.
S: Me too. Let's go.

Questions
1. What is Mr. Hawkes making?
   a. breakfast
   b. lunch
   c. dinner
2. Hencher thinks the police will catch them
   a. when they buy eggs and bacon.
   b. when they spend a lot of money.
   c. when they go to jail.

3. How did Sparrow and Hencher feel that morning?
   a. sad at first, then happy
   b. happy at first, then scared
   c. happy all morning
4. How long has Foley been missing?
   a. less than 24 hours
   b. 24 hours
   c. more than a day
5. What does Mr. Hawkes tell Sparrow and Hencher?
   a. They should ask for $100,000 for the horse.
   b. They should go to California.
   c. They should return the horse.
6. If someone sees them returning the horse, they will say they were
   a. going to California.
   b. stealing the horse.
   c. just borrowing Foley.

Part 2
Peg is in a bar, sitting by a window.

Peg: Bartender! Bring me another beer!
Bartender: You've already had' two, Peg. It's only 11:30.
P: So what! Who cares what time it is?
B: At this rate you'll be drunk by noon.
P: Who gets drunk on three beers? My father drinks six and acts perfectly normal.
B: Everybody's different. You're not used to drinking. But if you want it ... (sound of beer opening and being put down)
P: Thanks. I'm so worried. Foley's been stolen.
B: You can't be sure of that. Maybe he just got lost.
P: Expensive racehorses don't get lost! I pay Sparrow and Hencher to take care of him.
B: Well, it's not such a big loss. Foley just isn't as fast as Winni and Studley were. Some children grow up very different from their parents.
P: He'll get faster. He's very young. Say—that's my Foley!
B: Where?
P: In that truck! I've got to get him back! (sound of table and bottle falling)
B: You should call the police!
P: See you later!

Questions
1. What time is it?
   a. 11:30 A.M.
   b. noon
   c. 11:30 at night
2. How many beers did Peg order?
   a. 2
   b. 3
   c. 6
3. How many beers did she finish?
   a. 2
   b. 3
   c. 6
4. The bartender suggests that Foley was
   a. stolen.
   b. lost.
   c. borrowed.
5. The bartender says Foley was
   a. faster than Winni and Studley.
   b. not as fast as Winni and Studley.
   c. just like his parents.

6. When Peg jumps up, the bartender says she should
   a. not drink another beer.
   b. call the hospital.
   c. call the police.

CASE 6

Mr. Bilder has made an appointment to see Mr. Farmer and look at lot A.

Bilder: This looks like a fine piece of property, and lots of city people are moving out here. They're going to need movie theaters, laundromats, and shopping centers. This lot is perfect for a shopping center.

Farmer: Maybe it is, but I'd never allow it. I can't stand all that noise, the traffic, and I just don't like people very much. I'm happy here with my dogs and cows and sheep.

B: Then why did you sell the other lot?

F: I needed the money badly. Farming is hard work, and I'm not young anymore.

B: What did you sell the other lot for?

F: $10,000.

B: I'll pay you $50,000 for this one, if I can use it for commercial purposes.

F: That's a lot of money, but I have to say no.

B: $100,000.

F: $100,000! I'll take it! You can build anything you want.

F: Are you sure the noise won't bother you.

F: I'll put dollars in my ears if it does.

B: You must realize that I may have to cut down some of those trees that divide our properties.

F: Go ahead. With this much money, I can go to Florida.

B: One more thing—there aren't enough people here for two shopping centers. That young man Ayer has already started building one. He's going to lose a lot of money because of that, but you can't let him continue.

F: Don't worry. He can't. It's in the contract.

Questions
1. People are moving here from
   a. the suburbs.
   b. the city.
   c. the country.
2. According to Mr. Bilder, this is a perfect place for
   a. a laundromat.
   b. a theater.
   c. a shopping center.
3. Which animal doesn't Mr. Farmer have?
   a. a cat
   b. a dog
   c. a cow
4. Mr. Bilder's first offer for the land is
   a. $5,000.
   b. $50,000.
   c. $100,000.
5. If the noise becomes too great, Mr. Farmer will
   a. go to California.
   b. build anything.
   c. put dollars in his ears.

6. Which is most important to Mr. Farmer?
   a. sheep
   b. money
   c. noise
7. If Mr. Bilder cuts down a lot of trees, Mr. Farmer might
   a. go to Florida.
   b. go to jail.
   c. sell the trees.
8. Who is probably going to lose a lot of money?
   a. Mr. Bilder
   b. Mr. Farmer
   c. Ayer

CASE 7

Neville and E. Z. are at Jack's on the night of the accident.

E. Z.: How about one more beer, Neville?

Neville: OK. Just one. I shouldn't get too drunk. I have to ride home.

E: Bartender, bring me another whiskey, and one more beer for Neville.

Bartender: Don't you think you've had enough, E. Z.? You've already had five.

E: It's OK. I don't have to drive. Neville's giving me a ride home on his new bike.

B: Why don't you call a taxi? Neville's not supposed to drive.

E: Why is that?

N: I lost my driver's license last week for drunk driving.

E: Don't worry. It's only a couple of miles. You feel well enough to drive, don't you Neville?

N: I guess so. But I'm worried about you. Can you hold on?

E: Sure. (sound of glasses being put on table)

B: Here you are, boys. Drink up. The bar closes in twenty minutes.

Questions
1. Why doesn't Neville want to get too drunk?
   a. because he has to call a friend
   b. because the bar is closing
   c. because he has to ride home
2. What is Neville drinking?
   a. whiskey
   b. beer
   c. water
3. How many whiskeys has E. Z. already drunk?
   a. 5
   b. 1
   c. 11½
4. Why isn't Neville supposed to drive?
   a. because he lost his way
   b. because he lost his license for drunk driving
   c. because he lost his license after an accident
5. Is Neville sure about his ability to drive?
   a. Yes.
   b. No.
6. Whom is Neville worried about?
   a. the bartender
   b. E. Z. Rider
   c. Jack

7. When do E. Z. and Neville leave the bar?
   a. early
   b. exactly thirty minutes before closing time
   c. shortly before closing time

CASE 8

Part 1
Rick's boss is standing in the doorway of her office when Rick comes to work at the newspaper.
Rick: Good morning, boss.
Boss: Morning? It's afternoon! You're late again, Rick! You're supposed to be here at 8:30, and it's 12:30.
   R: I was writing a story about the fire on Main Street. I was watching the fire until five o'clock this morning. How can you expect me to come in at 8:30? I have to get some sleep!
   B: I'm sorry, but I have to let you go. You're finished! Clean out your desk and go home. You're not working for the *Daily News* anymore.
   R: That's not fair! You can't do that!
   B: I just did! (door slams)
   R: (banging on door) You'll be sorry for this! If I ever get my hands on you I'll break your neck!
   B: Take a walk! Go feed the birds!

Questions
1. What time did Rick arrive at work?
   a. 8:30
   b. 12:30
   c. 5:00
2. Where is Rick's boss standing?
   a. in her office
   b. on Main Street
   c. in the doorway
3. What was Rick watching?
   a. an accident
   b. a fight
   c. a fire
4. Rick thinks his boss's action is
   a. fair.
   b. not fair.
   c. not kind.
5. How does Rick feel when he leaves?
   a. angry
   b. sad
   c. fine

Part 2
Listen to the conversation between Martin Gardiner and his boss.
Boss: Martin! Put that cigarette out and get back to work!
Martin: I'm on my lunch hour.
   B: You only have a half hour for lunch. Now get to work. You're the laziest employee I have.
   M: It's too hot to work today.
   B: Then work on the pond. I heard there are holes developing in it. Someone might fall in and get hurt.
   M: Yeah, maybe a fish.

B: Who knows? Maybe a kid will be playing in the pond and hurt himself.
M: Nobody is supposed to go into the pond.
B: I know, but kids will be kids.
M: Those holes have been there for a year and nobody has got hurt. They can wait. I think I'll cut the grass.

Questions
1. What is Martin doing when his boss sees him?
   a. drinking
   b. smoking
   c. sleeping
2. How much time does Martin get for lunch?
   a. half an hour
   b. an hour
   c. an hour and a half
3. The boss thinks Martin is
   a. smart.
   b. kind.
   c. lazy.
4. How is the weather this day?
   a. hot
   b. cold
   c. rainy
5. How long have the holes been there?
   a. a week
   b. an hour
   c. a year
6. What is Martin going to do instead of fixing the holes?
   a. have lunch
   b. smoke a cigarette
   c. cut the grass

CASE 9

Listen to the telephone conversation between Ann and Dr. Young. Bill is in the TV room.
                    (sound of phone dialing)
Dr. Young: Hello. Dr. Young speaking.
       Ann: Hello, Doctor, this is Ann Moakley.
         Y: What can I do for you?
         A: I'm worried about Bill.
         Y: You shouldn't worry about him. He's making a lot of progress.
         A: It doesn't look that way from here. Last night he was so crazy I asked him to leave the house. Then he said he would kill me and the kids.
         Y: Now, now, take it easy. You know Bill. He'll say anything that comes to his head. He doesn't mean it. The problem is he's watched so much TV he acts like a person on the TV. You really can't believe what he says.
         A: Well, I'm still worried, but the doctor knows best. This is giving me a terrible headache.
         Y: Take two aspirin and get some rest.
         A: All right. Goodbye, Doctor.
         Y: Goodbye. (phone hangs up)
      Bill: Who was that?
         A: Oh, just a friend.

65

B: Who? Can't you tell me his name?

A: It's none of your business. Leave me alone. I've got a headache.

B: It was your ex-husband, wasn't it?

A: No. Let go of my arm! Ow! You're hurting me!

B: Where's my breakfast?

A: I didn't know you were up.

B: I've been watching TV for two hours already with no breakfast. The kids are hungry too. Now get in the kitchen before I kill you.

Questions

1. What did Ann ask Bill to do last night?
   a. make dinner
   b. watch TV
   c. leave the house
2. Last night, Bill said he would
   a. kill the kids.
   b. kill time.
   c. leave the house.
3. Is Dr. Young worried about Bill?
   a. Yes.
   b. No.
4. Does Dr. Young think Bill means what he says?
   a. Yes.
   b. No.
5. Does Ann have confidence in the doctor?
   a. Yes.
   b. No.
6. Ann tells Bill that the person she was talking to was
   a. Dr. Young.
   b. her ex-husband.
   c. a friend.
7. Who wants breakfast?
   a. Just Bill
   b. Ann
   c. Bill and the kids
8. What was Bill doing while Ann was on the phone?
   a. watching television
   b. sleeping
   c. having breakfast
9. Bill says that if Ann doesn't get into the kitchen he will
   a. hit her.
   b. be hungry.
   c. kill her.

CASE 10

Part 1

Listen to the conversation between Daphne and the circus manager.

Vendor: Peanuts! Popcorn! Candy! Hot dogs! Peanuts!

Circus
Manager: Get your tickets here for the big show! See the elephants! The lions!

Daphne: Excuse me, are you the circus manager?

CM: Yes. Who are you?

D: My name's Daphne. I'm looking for Mr. Gray, the knife-thrower.

CM: Oh, God, not another one!

D: Another what?

CM: Mr. Gray doesn't work for us anymore. He hasn't worked for us for two years.

D: That's impossible. I practiced the knife-throwing trick with him this afternoon. He wants me to be his assistant.

CM: You and a dozen other girls. Every couple of months he fools another girl like you. The guy is crazy. That's one of the reasons he doesn't work for us anymore.

D: Crazy?

CM: Yes. And his eyes were getting bad. Was he wearing his glasses this afternoon?

D: No.

CM: You were lucky then.

D: Oh my God! You mean he could have killed me? Oh! Oh! (she faints)

CM: Young lady, are you all right? Somebody get a doctor, quick!

Questions

1. What is the circus manager selling?
   a. peanuts
   b. tickets
   c. popcorn
2. Which animal cannot be seen at the circus?
   a. a dog
   b. a lion
   c. an elephant
3. How many girls has Mr. Gray fooled?
   a. 2
   b. about 13
   c. about 25
4. How were Mr. Gray's eyes?
   a. getting bad
   b. getting better
   c. excellent
5. What does the circus manager call for?
   a. a police officer
   b. a hot dog
   c. a doctor

Part 2

Listen to the conversation between Daphne and her mother.

Mrs. Disney: Daphne, how are you?

Daphne: Not bad. But I keep having dreams about a knife going through my head.

Mrs. D: Are you eating well?

D: Not really. You know how hospital food is. Where's daddy?

Mrs. D: You know how busy your father is. He is in court today. He might be able to come see you next week.

D: Oh. And how are you, mother?

Mrs. D: I'm angry, that's how I am. And I'm more angry with you than with that crazy man Gray. How could you be so stupid? If a magician said to you, "I can cut you in half and it won't hurt," would you let him do it?

D: Of course not. But Mr. Gray seemed like a nice man.

Mrs. D: He'll pay for this! I promise! He made a big mistake fooling with the daughter of a lawyer.

1. Where is Daphne?
   a. in the hospital
   b. at home
   c. at college
2. What does Daphne dream about?
   a. a magician
   b. bad food
   c. a knife
3. Whom is Mrs. Disney more angry with, her daughter or Zane Gray?
   a. her daughter
   b. Zane Gray
4. Where is Mr. Disney?
   a. at a baseball game
   b. at home
   c. in court
5. When might Mr. Disney be able to see Daphne?
   a. today
   b. next Friday
   c. next week
6. What is Mr. Disney's profession?
   a. doctor
   b. lawyer
   c. teacher

1. What is the doctor doing when the police captain first speaks to him?
   a. making an examination
   b. eating
   c. writing a report
2. How does the captain feel about this place?
   a. It makes him sick.
   b. He likes the view.
   c. The service is good, but the food is not.
3. What did Ruth die of?
   a. scarlet fever
   b. a fever
   c. medicine
4. Did the medicine make Ruth worse?
   a. Yes.
   b. No.
   c. Maybe.
5. What is the captain's opinion of the doctor?
   a. He is excellent.
   b. He is only good for stomach problems.
   c. He is not very good.
6. What did this doctor and Dr. Payne do together?
   a. play golf
   b. go to restaurants
   c. do operations
7. Why shouldn't the police captain call this doctor if he had a fever?
   a. because the doctor plays golf on Wednesdays
   b. because the doctor only does check-ups
   c. because the doctor only examines dead bodies

CASE 11

The police ordered an examination of Ruth's body to see what she died of. Listen to this conversation between the police captain and the doctor who did the examination.

Police
Captain: Well, doctor, did you finish the examination?

Doctor: Could you hand me the salt, please, and the butter?

PC: How can you eat with all these dead bodies around?

D: You get used to it. And it doesn't bother them. At least, they haven't said anything about it.

PC: This place makes me sick to my stomach.

D: Then please don't eat anything, captain. Now, as for the results, Ruth died of a fever.

PC: Scarlet fever?

D: No. Definitely not. It was a fever, but not scarlet fever. Dr. Payne was wrong.

PC: What a doctor! If that was my kid, I think I'd shoot him too. So, what about the medicine Dr. Payne gave her for scarlet fever? Did it have any effect?

D: It's impossible to tell. Maybe the medicine helped a little, and maybe it made the fever worse.

PC: You mean you don't know?

D: Right.

PC: I hope I never get sick. And if I do, I won't call you.

D: If someone calls me it will be too late. All I do is examine dead bodies. (laughs) Poor Dr. Payne. It's really too bad. He was my golf partner every Wednesday, you know.

CASE 12

Listen to the conversation. Police Captain DesMaisons picks up Jack Henderson and Mr. Mathers in his car. He is taking them to the police station for questioning.

Police
Captain: Jack, what did you do on the day Karen died?

Jack: Not much. I went shopping in the morning.

PC: What did you buy?

J: A pound of hamburger and a half gallon of ice cream.

PC: What kind?

J: Chocolate.

PC: Do you like chocolate?

J: Uh, yes, it's my favorite.

PC: Weren't you afraid your wife might eat it all and die?

J: No, Captain. She's not a child.

PC: Did you buy anything else?

J: Not that I can think of.

PC: We found a sales receipt. There were three prices on it. You didn't buy a box of chocolates, did you?

J: No. Now I remember—I bought some cigarettes too.

PC: I have the receipt right here. Let's see— $2.19, that must be the ice cream; $1.75, that must be the hamburger. The last is $3.50. That's five packs of cigarettes. Doesn't that seem strange to you? I can understand one or two packs of cigarettes, or a carton of ten, but I don't understand five packs.

J: That's all I could afford. I didn't have enough money for a carton.

PC: It's strange, though. $3.50 is the average price of a box of chocolates. (loud cough)

J: You ought to let Mathers take a look at you, Captain. He used to be a doctor.

Mathers: Shut up! Mind your own business.

PC: Jack, what did you do after you went shopping?

J: I went to the race track. It was my lucky day, Captain. I won $5,000. It's too bad Karen isn't here to share it.

PC: Did she ever try to commit suicide before?

J: Uh . . . yes. Last year, in May.

PC: Who was her doctor then?

J: It was M . . . uh, I can't remember.

PC: Jack, do you speak French?

J: No.

PC: How about you, Mathers?

M: Just English, Captain.

PC: Jack, can you type?

J: No.

PC: Mathers?

M: Of course.

PC: God, I'm hungry. I think I'll stop at that ice cream place. Can I get you boys anything?

M: I'll take a vanilla ice cream cone.

J: Me too. I love vanilla.

Questions

1. Where are they?
   a. in the police station
   b. in the train station
   c. in a car
2. What is Jack's favorite kind of ice cream?
   a. chocolate
   b. strawberry
   c. banana
3. What kind of ice cream did Jack order?
   a. chocolate
   b. strawberry
   c. vanilla
4. How many packs are in a carton of cigarettes?
   a. 2
   b. 5
   c. 10
5. Why didn't Jack buy a carton of cigarettes?
   a. He only wanted two packs.
   b. He only wanted six packs.
   c. He didn't have enough money for a carton.
6. What was Mathers's previous profession?
   a. president
   b. doctor
   c. lawyer
7. Does Mathers want the captain to know he was a doctor?
   a. Yes.
   b. No.
   c. He doesn't care.
8. How much did Jack win at the track?
   a. $10
   b. $5,000
   c. $15,000
9. When did Karen try to commit suicide before?
   a. 1 year ago
   b. 2 years ago
   c. 5 years ago
10. Does either Jack or Mathers speak French?
    a. Both do.
    b. Neither one does.
    c. One of them does.
11. Who can type?
    a. Jack
    b. Mathers
    c. DesMaisons

ANSWER KEY

CASE 1

A. *Vocabulary:*
1) d   3) c
2) b   4) a
B. *Cumulative Vocabulary Exercise:* judge, jury, verdict, furious, twist, temper, illegal
C. *General Exercises:*
1. *Sequencing:* 6, 3, 4, 5, 1, 2
2. *Fill In:*
   1) three   3) two
   2) thirteen   4) pocket
E. *Pre-trial Questions:*
1) D   6) D
2) D   7) D
3) D   8) P
4) D   9) P
5) P
F. *Listening Comprehension:*
1) b   5) b
2) c   6) b
3) c   7) d
4) c

CASE 2

A. *Vocabulary:*
1) d   3) d
2) c   4) e
B. *Cumulative Vocabulary Exercise:* thief, fired, ghost, witnessed, thief, ruined
C. *General Exercises:*
1) D   7) C
2) H   8) L
3) J   9) B
4) A   10) E
5) I   11) F
6) G   12) K
E. *Pre-trial Questions:*
1) Z   6) Z
2) L   7) L
3) Z   8) Z
4) Z   9) L
5) Z
F. *Listening Comprehension:*
1) b   5) c
2) a   6) b
3) b   7) c
4) c   8) b

CASE 3

A. *Vocabulary:*
1) a   6) a
2) d   7) c
3) c   8) d
4) e   9) b
5) c   10) a
B. *Cumulative Vocabulary Exercise:* teenage, woods, illegal, arrest, crime, mercy, sentence, climbed, irrelevant, sour
F. *Listening Comprehension:*
1) c   4) b
2) b   5) a
3) a   6) c

CASE 4

A. *Vocabulary:*
1) b   6) d
2) c   7) c
3) d   8) d
4) d   9) a
5) a   10) a
B. *Cumulative Vocabulary Exercise:* furious, assumption, wealthy, greed, contract, century
C. *General Exercises:*
1) a   3) b
2) e   4) b
F. *Listening Comprehension:*
1) b   4) a
2) c   5) b
3) b

CASE 5

A. *Vocabulary:*
1) race   6) dead end
2) retired   7) chased
3) champion   8) injured
4) colt   9) scar
5) stables

B. *Cumulative Vocabulary Exercise:* robbed, thieves, stole, robbed, thefts, thieves, thief, steals, theft
C. *General Exercises:* Match:
1—b 4—a
2—e 5—c
3—d
F. *Listening Comprehension:*
Part 1:
1) a 4) a
2) b 5) c
3) b 6) c
Part 2:
1) a 4) b
2) b 5) b
3) a 6) c

3. *Fill in . . . Preposition:*
a) on, of, off, of d) in, of, of
b) over, on, of e) into, to
c) for, from, to f) after
4. *Rewrite:*
a) She was driving her car one night. (*or:* One night she was driving her car.)
b) She always has a drink at a bar after she finishes work.
c) I called a taxi cab for my friend.
d) You should never walk alone in the city after dark.
e) I think Neville is the person at fault. (*or:* I think that the person at fault is Neville.)
f) Neville has only himself to blame for his drinking problem.
F. *Listening Comprehension:*
1) c 5) b
2) b 6) b
3) a 7) c
4) b

CASE 6

A. *Vocabulary:*
1) lots 5) construction
2) boundary 6) shopping center
3) commercial 7) situation
4) purposes 8) bother
B. *Cumulative Vocabulary Exercise:* constructing, track, races, bother, commercial, illegal, retire
E. *Pre-trial Questions:*
1) Ayer 4) Ayer
2) Ayer 5) Farmer
3) Farmer
F. *Listening Comprehension:*
1) b 5) c
2) c 6) b
3) a 7) a
4) b 8) c

CASE 7

A. *Vocabulary:*
1) d 3) d
2) e
B. *Cumulative Vocabulary Exercise:* champion, races, chased, punishing, instantly, racing
C. *General Exercises:*
1. *Fill in . . . Fact:*
a) two miles d) medical examiner's
b) Jack's e) fence
c) Honda f) telephone pole
2. *Sequencing:*
a) 8 f) 5
b) 7 g) 2
c) 6 h) 9
d) 1 i) 3
e) 4

CASE 8

A. *Vocabulary:*
1) c 5) a
2) d 6) b
3) a 7) d
4) d 8) c
B. *Cumulative Vocabulary Exercise:* rescue, drowning, Fortunately, races, judge, lane, accused, stealing, ignore, accusation, furious, sue
C. *General Exercises:*
1. *Fill in:*
a) enough e) too
b) too f) very
c) very, enough g) enough
d) enough
2. *Rewrite . . . Sentences:*
a) Martin let holes develop.
b) I consider the one who pushed him more guilty.
c) What I think is that it was an accident.
d) She should never leave the boy alone.
e) Deke let the boy swim by himself and returned to his conversation with Cher.
f) He read about it in the newspaper.
3. *Rewrite . . . Words:*
a) It was a ten-foot snake.
b) He was a ten-year-old boy.
c) He got a new three-year contract.
d) The bullet left a one-inch-deep hole in his chest.
F. *Listening Comprehension:*
Part 1:
1) b 4) b
2) c 5) a
3) c
Part 2:
1) b 4) a
2) a 5) c
3) c 6) c

CASE 9

A. *Vocabulary:*
1) b   5) c
2) e   6) a
3) d   7) b
4) e

B. *Cumulative Vocabulary Exercise:* rifle, stealing, fired, shots, thief, scared, bullet, fortunately, arrested, trial, defense, psychiatrist

C. *General Exercises:*
1. *Sequencing:*
a) 2   d) 1
b) 5   e) 4
c) 3   f) 6
2. *Fill in:*
a) of   e) of, at
b) In, in   f) with
c) in, of   g) At, on, of
d) from

F. *Listening Comprehension:*
1) c   6) c
2) a   7) c
3) b   8) a
4) b   9) c
5) a

B. *Cumulative Vocabulary Exercise:* unloaded, basement, stable, load, warn, court, shooting, witness, prosecution, beating

C. *General Exercises:*
1. *Sequencing:*
a) 1   f) 5
b) 9   g) 6
c) 3   h) 4
d) 8   i) 7
e) 2
2. *Fill in . . . Fact:*
1) wedding   6) shipyard, New Jersey
2) Italian-American   7) sixteen
3) six   8) 11:00 P.M.
4) two   9) August
5) two
3. *Fill in . . . Preposition:*
1) in, for   5) for
2) in, for   6) of, to (with)
3) as, in (at), in 7) up, as, as
4) at, after   8) with, at
4. *True or False:*
1) T   6) F
2) F   7) F
3) F   8) F
4) F   9) F
5) T

F. *Listening Comprehension:*
1) b   5) c
2) a   6) a
3) b   7) c
4) c

CASE 10

A. *Vocabulary:*
1) a   5) e
2) e   6) a
3) b   7) e
4) c   8) c

B. *Cumulative Vocabulary Exercise:* fail, Introduction, assistance, semester, woods, lot, boundary, shock, rescued, cowboy

F. *Listening Comprehension:*
Part 1:
1) b   4) a
2) a   5) c
3) b
Part 2:
1) a   4) c
2) c   5) c
3) a   6) b

CASE 11

A. *Vocabulary:*
1) d   6) c
2) a   7) b
3) a   8) b
4) e   9) d
5) e

CASE 12

A. *Vocabulary:*
1) d   7) b
2) c   8) d
3) c   9) a
4) a   10) a
5) d   11) c
6) d

B. *Cumulative Vocabulary Exercise:* diary, retirement, books, diabetes, quit, insulin, crazy, discriminate, guilty, revenge

C. *General Exercises:*
1. *True or False:*
1) F   8) F
2) F   9) F
3) F   10) F
4) T   11) T
5) F   12) F
6) T   13) T
7) F
2. tell, newspapers, paid its taxes.
3. I have a terrible secret.

F. *Listening Comprehension:*
1) c   7) b
2) a   8) b
3) c   9) a
4) c   10) b
5) c   11) b
6) b

PROCEEDINGS OF THE INSTITUTION OF CIVIL ENGINEERS

The Channel Tunnel

SUPPLEMENT TO CIVIL ENGINEERING

CONTENTS

Part 1: Tunnels

*The papers published in this special
issue will be presented at a meeting
at the Institution of Civil Engineers
on 2 February 1993, commencing
at 2.00 p.m.*

*Written discussion will close on 15
February 1993.*

*Channel Tunnel project:
completed running tunnel.*

Proc. Instn Civ.
Engrs, Civ. Engng,
Channel Tunnel,
Part 1: Tunnels,
1992, 2–5

Paper 9984

Concept, reality and expectations

A. F. Gueterbock, OBE, MA, MICE

After many years of debate, the Channel Tunnel will shortly become a reality for the road and rail customers who have waited so long to see Britain's transport network connected to the Continent. To achieve this, design and construction teams have worked together in a unique bi-national project, often under an international media spotlight. This Paper introduces seven other papers, covering the civil engineering management, design and construction of the UK section of the Channel Tunnel. The design and construction of the UK terminal works and the wider transport infrastructure issues will be the subject of future papers.

2. The papers have been written by those involved in the project, from the development of the concept started by Sir Harold Harding in the late 1960s to completion of construction in 1992/3.

3. Paper 9962 describes the organization and management roles of the parties to a fast track project, which was a requirement of private sector funding. The roles of other external bodies, such as the Intergovernmental Commission, are also discussed.

4. Paper 9963 provides an overview of the design and construction of the tunnels and associated structures. It offers an insight into the organization and interfaces of the design and construction teams; it also describes some of the associated problems and the need to integrate civil, mechanical and electrical works into a railway transport system.

5. Paper 9932 sets out the geology of the Dover Straits, the geological and operational reasons for the choice of alignment, and the problems and successes of the underground survey operations.

6. Perhaps the most important aspect of the construction and the barometer of progress, from the financing point of view, was the successful development and operation of the tunnel boring machines (TBMs). Paper 9964 describes these, as well as their installation and commissioning.

7. There are many other underground structures which could not be built by TBMs. Some, such as the preliminary underground works at Shakespeare Cliff, were required for TBM erection. Others, such as the undersea crossover and Castle Hill tunnel, are an integral part of the permanent works. Paper 9933 describes in detail such underground structures.

8. Paper 9934 provides a fascinating account of a vital part of the fast track approach, the development and operation of the Lower Shakespeare site to supply all the materials required for the underground works as well as a place for spoil disposal, creating the essential working area for the construction works themselves.

9. Lastly, Paper 10045 covers the design and manufacture of the 450 000 precast concrete tunnel linings to a very exacting specification, as well as the procurement of SGI and other linings.

History

10. Ideas and proposals for a Channel Tunnel or another type of fixed link go back to Napoleon's time, when French mining engineer Albert Mathieu-Favier put forward an ambitious proposal for a tunnel ventilated by chimneys rising above the waves and lit by gas lamps.

11. In the mid 19th century the French engineer Thomé de Gamond spent 40 years attempting to find practical solutions, and risked his life to conduct the first systematic geological survey of the sea-bed when he dived from a boat to take samples with weights on his feet, pig bladder floats, ears full of fat and fabric, and a mouth full of olive oil for protection against the water pressure. Just to round things off, he was apparently attacked by giant conger eels during the dive.

12. In the 1880s, construction of a Channel Tunnel actually started at Shakespeare Cliff and Sangatte. The promoter, Sir Edward Watkins of the London and South Eastern Railway Company, took many parties of Members of Parliament and other VIPs for dinner underground to demonstrate how well the tunnelling was going. Nevertheless, it was stopped because the British military authorities feared invasion.

13. A further attempt was made in the 1920s, and a trial bore using the Whitaker Machine was undertaken at the Folkestone Warren. This machine was rescued last year from the cliffs and is now refurbished and in working order at the Eurotunnel Exhibition Centre in Folkestone.

14. In 1955 Minister of Defence Harold Macmillan finally announced that a tunnel would no longer be a threat to national security. Discussions, plans and schemes started again, leading to the construction of a bored rail tunnel starting at Shakespeare Cliff and Sangatte in 1974. Once more, it was stopped by the British government, which appeared to lack commitment to the financial guarantees considered necessary, as well as to the high cost of the high-speed line to London. Unsurprisingly,

A. F. Gueterbock,
Public Affairs
Manager,
Eurotunnel

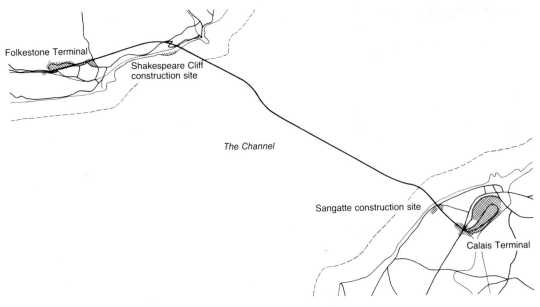

Fig. 1. Tunnel route

this decision caused much annoyance to the French authorities, and it took great efforts in the following decade to rebuild the necessary trust between the two governments.

15.   In 1978 British Rail and SNCF proposed a 6 metre diameter single-track tunnel, generally known as the Mousehole, attempting to demonstrate that a very much cheaper solution could still bring substantial benefits to the railways and to Britain.

16.   However, the British government in particular felt that any fixed link should be promoted and financed in the private sector, and so in 1980 it invited ideas and proposals for fixed links to be constructed and operated in the private sector. Following various reports from the Government, banks and others, formal invitations to bid were issued in 1985, and on 20 January 1986 the Channel Tunnel scheme (as proposed by Eurotunnel, The Channel Tunnel Group/Francemanche) of a rail shuttle service for road vehicles with provision for through trains, was accepted. The route of the successful tunnel scheme is shown in Fig. 1 and the general layout of the tunnel system is shown in Fig. 2.

The need for a tunnel

17.   Many sound transport reasons have been put forward over the years for constructing a Channel Tunnel but, until recently, defence reasons for not constructing it always prevailed.

18.   More recently, as Britain joined the European Community and foresaw the approaching Single Market, it became evident that better and alternative means of communication were necessary to cope with the traffic growth, and to provide greater convenience as well as greater speed.

19.   Cross-channel traffic has virtually doubled in the last 20 years, reflecting Britain's increased trade with the Continent and a greater demand for travel and movement of freight. This is partly explained by the fact that over 60% of Britain's trade is now with continental Europe.

20.   Beyond these bare statistics, however, is the realization that both speed and reliability are important for freight deliveries, and if these cannot be guaranteed many multinational firms may reconsider the location of their manufacturing or distribution outlets.

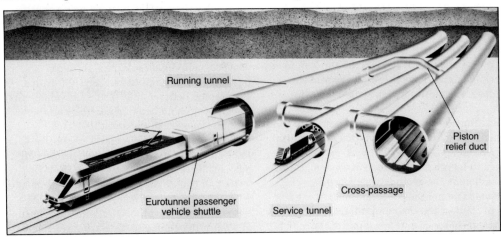

Fig. 2. Perspective of three-tunnel system

21. Similarly, the inconvenience and delay in changing one's mode of transport is something which is becoming less acceptable to the increasing number of people travelling for business and pleasure. In order to survive in today's markets, freight operators demand reliability, speed and the minimum of inconvenience.

22. Thus, the potential for the Channel Tunnel is to create an alternative competitive link between the road and motorway systems of the UK and France, competing in this case with the ferries. Of equal importance is the competition between the through-rail services, planned to operate between London, Paris and Brussels in about 3 hours, and the air services. Taking into account the delays and the travel from city centre to city centre, there is little doubt that the rail passenger services will be not only quicker but generally much more comfortable and reliable.

The tunnel becomes a reality

23. With the Channel Tunnel chosen, the signing of a treaty between Britain and France completed, and a concession agreement between Eurotunnel and the two governments made, a great deal of work had to be done before final approval to construct the tunnel was obtained.

24. In Britain a Parliamentary bill needed to be promoted and subjected to Select Committee hearings in both Houses of Parliament.

25. In France the process of receiving approval for the construction was less onerous, and the employment possibilities were generally welcomed in the Region Nord Pas de Calais.

26. Over and above all this was the need to demonstrate that, before permission was given for construction to start, sufficient finance had been raised to complete construction. At the very beginning, the original banker/contractor group was separated into client, contractor and lending banks and, inevitably, there was severe friction between these various organizations, all of whom wished to see the project succeed but often on rather different terms.

27. Despite the very demanding examinations in Parliament, Royal Assent to the Channel Tunnel Bill was received in July 1987, and by the end of November 1987 the loan agreements and the main equity issue were complete so that formal construction could start.

28. During the ensuing excitement over tunnel boring, the massive volumes of material supply and spoil disposal, it was easy for people to forget that a transport system, rather than a hole in the ground, was being constructed. The purpose of this transport system was to connect the road and rail networks of Britain and France and to fill one of the major missing links in the European transport system.

The tunnel system as a business

29. The Channel Tunnel scheme was developed in response to government guidelines stating that it must be financed in the private sector with no government assistance or guarantees.

30. Those developing the twin-bored rail tunnel that was to become the Eurotunnel scheme were therefore conscious that, if their scheme was to be as attractive as possible to banks and investors, it had to meet the following criteria

(a) minimum capital cost and minimum risk of overrun;
(b) minimum operating cost;
(c) maximum traffic revenue.

31. The successful scheme used more tried and tested techniques than those for the other competing schemes. 7·6 metre diameter running tunnels could be more easily constructed at speed than the 13 or 14 metre diameter tunnel needed for a two-lane road tunnel.

32. For any private-sector-funded scheme short construction time is as important as construction costs, owing to the very high interest charges which would occur in the event of serious construction delays. Similar comments can be made in respect of any rolling stock which, in the case of the Channel Tunnel, is necessary to take the traffic through the tunnel.

33. The decision to carry road transport on rail wagons was made for very sound technical and driver-psychology reasons. This virtually decided the tunnel configuration and diameter.

34. However, as can be seen from the prospectus issued by Eurotunnel in 1987, the most critical factor in its successful financing was the likely revenue. Customers must want to use it, and the service must therefore be attractive, convenient, reliable and competitively priced. It must also have the capacity to deal with peak demands.

35. A rail-operated system, using shuttle trains for road vehicles, provided capacity, reliability and flexibility, and gained Eurotunnel revenue from the widest possible sources: cars, coaches, motor cycles, lorries, through-passenger trains and through-freight trains. Eurotunnel believes that the motor vehicle customers want a no-booking service, and therefore the capacity of the shuttle service, combined with any differential pricing, needs to be sufficient to cope with such periods of peak demand.

36. Of course the most important element of all is the total cross-Channel market for the types of traffic, combined with an assessment of what proportion of each market segment is

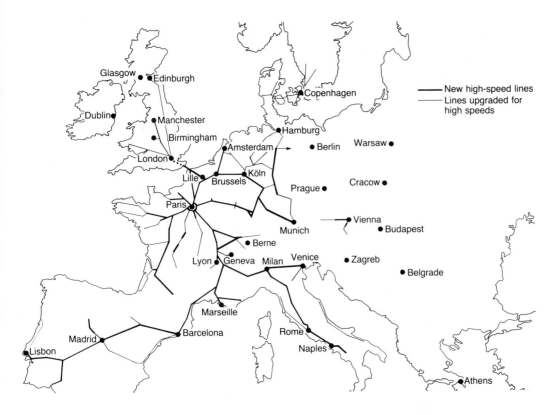

Fig. 3. European
high-speed rail system
(proposed for year
2000)

likely to be diverted to the tunnel. As these
forecasts of traffic and the resulting revenue
were so vital to Eurotunnel's financing, independent traffic and revenue consultants were
employed to provide this information on a
regular and consistent basis. All parties have
been helped in this by a generally accepted
view that the total cross-Channel traffic is
likely to double in the next 15 to 20 years.

37.   Eurotunnel's objective from the start
was therefore to create a transport link across
the Channel bringing the maximum benefit to
its shareholders. The scheme chosen by the two
governments as that most likely to achieve this
was the bored rail tunnel.

38.   Whereas Eurotunnel was awarded a
concession to operate this tunnel for a period of
55 years, there is a similar obligation on the
British and French governments to provide rea-

sonable infrastructure connecting to the tunnel.

39.   There is, therefore, continuing debate,
especially in the UK, about whether this reasonable infrastructure has been achieved or even
planned, and how the Channel Tunnel will fit
into the long-term strategic transport policy or
plan published by the European Commission
and certain Member States, with the notable
exception of the UK.

40.   Some idea of such a plan for rail passenger is shown in Fig. 3: the European high-
speed rail system, as proposed by the
Commission of European Railways for the year
2000.

41.   Although Eurotunnel's initial plans are
to have a roughly equal allocation of traffic
capacity between road and rail, it will be interesting to see, in 10 or 20 years' time, whether
this is borne out in practice.

Proc. Instn Civ. Engrs, Civ. Engng, Channel Tunnel Part 1: Tunnels 1992, 6–17

Paper 9962

B. Patterson, UK Works Director, Eurotunnel

P. L. Allwood, Construction Director, Translink Joint Venture

P. Middleton, Maître d'Oeuvre, Eurotunnel

C. J. Kirkland, Technical Director, Eurotunnel

Management of the project

B. Patterson, BScEng, MIMechE, P. L. Allwood, FICE, P. Middleton, FIMechE and C. J. Kirkland, OBE, FEng, FICE

Introduction

The Channel Tunnel project is the creation of a fixed link transportation system between the UK and France. It is now well on the way to completion, thus bringing to fruition a scheme which has been talked about for 200 years.

2. The project is huge by any standard, and a number of other key factors impacted on the participants, some of which were: technical and logistical complexity, bi-nationality, private funding, time-scale, creation of an owner/operating company, and external influences.

3. This Paper shows how the organizations of the client, the contractor and the Maître d'Oeuvre evolved to enable each to fulfil its obligations. Before going on to these, it is worth summarizing the organization of the overall project which has now been in place since early 1988.

4. Eurotunnel (ET) is the owner and operator, financed by the shareholers and the funding banks. The governments of the UK and France are represented through the Inter-Governmental Commission, which is advised by the Safety Authority (SA) and the Maître d'Oeuvre (MdO). The contractor Transmanche Link (TML), consisting of two groups, each of five major UK and French contractors, respectively, are responsible for the design, construction, testing and commissioning of the project. The Maître d'Oeuvre (MdO) is the independent technical auditor reporting to the banks and the Inter-Governmental Commission, and advising the parties on matters affecting the project. The Joint Consents Team is responsible for managing the relationship with local government.

5. Each of the links between the entities in Fig. 1 represents a formal relationship entered into by the parties involved, and these are illustrated in Fig. 2.

Treaty

6. The treaty signed at Canterbury on 12 February 1986 set up the political framework between the UK and France within which the Tunnel has been constructed and will be operated. It addressed for example the question of jurisdiction, national boundaries and government involvement, and it set down in broad terms the role of the IGC and the SA.

DUP and Channel Tunnel Act

7. These two pieces of legislation enacted in France and the UK, respectively, established the political framework in each country within which the project was constructed and operated. They addressed for example, the topographical limits of the project and the planning regime applicable. They also described matters related to but outside the limits of the project itself, such as road and rail infrastructure.

Concession

8. This document described, among other things, the role and duties of Eurotunnel as concessionaire in relation to the IGC as representatives of the governments, and it also established the role and duties of the MdO in this context.

Loan agreement

9. This document described the relationship between Eurotunnel and the banks, setting down the conditions of the loan and the rights and responsibilities of each of the parties. Reference was made to the MdO role in respect of the banks.

Construction contract

10. This is the contract under which Transmanche Link (TML) design, construct, test and commission the fixed link transportation system, and Eurotunnel (ET) accept and operate it.

MdO contract

11. This contract exists between Eurotunnel and the MdO and describes the role of the MdO as technical auditor and independent advisor to the project, the IGC/SA and the banks.

12. It is now helpful to summarize the major features of the construction contract which have impacted on the development of the organizational structure of the client, the contractor and the MdO.

13. Under the conditions attached to the invitation from the two governments, the financing proposals had to be demonstrably 'robust'. They therefore wished to have as much certainty in the cost estimation as was possible, reinforcing the need for the description of works to be as clear as practicable.

14. The invitation also required the contract form to be included with the proposal, thus setting the contractor group and the client a seven month timetable for the contract negotiations—a task that proved impossible. Discussions on the form of the contract began with the FIDIC conditions, with the intention of capturing their familiarity and then adjusting

for the particularities of the project. These particularities quickly swamped the FIDIC document.

15. The contract needed to recognize three distinct divisions, relating to the sharing of risk. The tunnel design and construction was to be undertaken on a 'target cost' basis. The design and construction of the two terminals and their links to the national road and rail networks, and all fixed equipment throughout the project, was to be priced as a lump sum. The railway rolling stock for the project was covered by what became known as procurement items for which cost provision was at the employer's risk—a provisional sum.

16. The contract sum was to be subject to inflation in accordance with the usual national procedures. This led to some complex negotiations in respect of the target works, where two sets of actual costs have to be 'de-escalated' back to 1985 prices for comparison with 1985 target cost.

17. So far as payment is concerned, it is accepted that the employer should bear the responsibility for providing the funding for the execution of the works. This led to an acceptance that each month's work should be 'forward-funded', to avoid the need for the contractor to fund the work while payments in arrears were agreed. Some very, detailed discussions ensued in an effort to define a method for forecasting future cash requirements and determining the extent to which funds provided had been expended. A disputes panel was created, comprising three independent experts (not arbitrators). Panel decisions are not binding, but unanimous decisions must be observed by both parties unless overturned in arbitration.

18. The huge question of project definition and quality specification also had to be

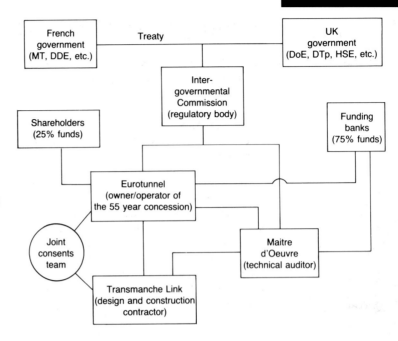

addressed. Although the project outline inherited from the 1974 scheme was fairly well understood, very little detail of the structures and operational control equipment had been developed. The descriptive schedules were therefore largely developed as a specification for performance, related to the forecast traffic.

19. A further complication was provided by the fact that the proposal contained no more than an outline of a scheme of tunnels and terminal arrangements which the proposers were able to demonstrate ought to work. Thus the contract called not simply for the working up of the proposal, but for the optimization of all aspects of the scheme. However it could not include state-of-the-art technology, since a price had to be agreed seven years before completion was due and the contractor was not prepared to

Fig. 1. Project organization

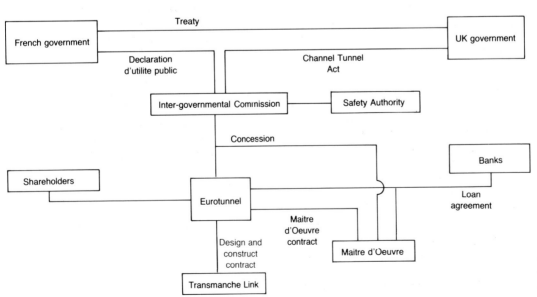

Fig. 2. Project organization with links shown

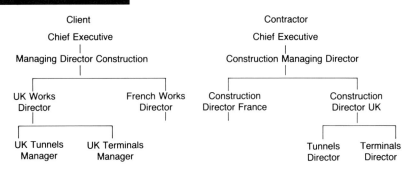

```
         Client                              Contractor
      Chief Executive                     Chief Executive
            |                                    |
  Managing Director Construction    Construction Managing Director
            |                                    |
    ┌───────┴───────┐                  ┌─────────┴─────────┐
  UK Works      French Works      Construction         Construction
  Director        Director        Director France       Director UK
    |               |                  |                    |
 ┌──┴──┐                                              ┌──────┴──────┐
UK Tunnels   UK Terminals                          Tunnels     Terminals
 Manager      Manager                              Director      Director
```

Fig. 3. UK construction management teams

give an overall warranty of fitness for purpose on completion. Acceptance criteria therefore had to be devised, which could be clearly tested, and some scale of performance set which would allow for some shortfall to be negotiated, rather than fixing a single absolute requirement.

20. Since the signing of the initial contract in 1986, two major re-negotiations have taken place, between ET and TML to resolve difficulties. These negotiations were supported by the banks as interested parties.

Role of the client

Creating a client

21. It is very unusual indeed that a major project is not initiated by the entity which will eventually own and operate the resulting facility.

22. For the Channel Tunnel project the process started with an open invitation from two governments for proposals for a Fixed Link which in turn motivated a group of construction contractors and banks to get together to submit a proposal. Neither the contractors nor the banks wished to be the client; their interests were shorter-term, but of course they recognized that a client would be necessary. It was foreseen that the client would be the recipient and owner of the concessions granted by the two governments, under which the link would be financed, constructed and operated for 55 years. This client would thus be responsible for procuring the finance, awarding the construction contract and managing it.

23. The proposal to the two governments was entirely prepared by the contractor—banking group, though it was submitted in the names of the Channel Tunnel Group and France Manche, the two companies which later formed the client Eurotunnel. These two companies, at the time of the proposal in November 1985, had been registered but only as a formality. They transacted 'only such business as was necessary to ensure the legality of the proposal', and their founder shareholders were the contractors and the banks. At this time, apart from an independent chairman and one other, their staff

were entirely drawn from the contractors and the banks. This staff was advised and assisted by Atkins and Halcrow in the UK and Setec in France, who were appointed technical auditors at this time.

24. As soon as it was realized, shortly after the submission of the proposal, that there was a very strong possibility that it would be successful, the need for an established client became urgent. The problem was money. Neither the banks nor the contractors wished to put more money into the formation of the client than was strictly necessary, but this client had to be capable of accepting the concession from the governments, of entering into a very large construction contract with the same contractors who now provided most of its staff, and of negotiating enormous commitment from the banks to finance the project and the company's own development.

25. Unfortunately, the time when the formation of the client became urgent coincided with the time when the contractors and the banks wished to withdraw some of their key staff who had worked in the proposal teams, to get on with the business of preparing for the start of the construction and financing.

Evolution of client capability

26. As has been stated earlier, at the outset the client was not much more than a name on a page, and almost all of the staff were provided by the contractors and the banks.

27. The invitation from the governments required the client to employ a Maître d'Oeuvre (overall supervising authority) during the construction phase. The proposer decided, however, to appoint his Maître d'Oeuvre before submission in order to give added strength to his proposal. This proved to be a good decision, not only for the proposal, but because the embryo client had the advice and support of the MdO through the early days of the project. The problem with this arrangement soon became apparent. As the contractor's design programme accelerated, more and more decisions or agreements were required of the client, and these were being formulated to an ever greater extent by the MdO teams. This was unacceptable, as it conflicted with the essential independent role required of the MdO.

28. The situation was resolved by the formation of a Project Implementation Division (PID) within Eurotunnel. At the peak of activity the client's Project Implementation Division numbered almost 400 persons, managing and administering the contract.

The organization and operation of PID

29. At the time the PID was formed an important and significant decision was taken to

structure the PID to match the contractor's organization. An example of this is the UK construction management teams, which are structured as shown in Fig. 3. As a result, there have always been clearly identified interfaces and communication links between all management levels within the PID and the contractor's organization.

30. The current structure of Eurotunnel's Project Implementation Division is as shown in Fig. 4. In general terms, Transportation Systems Engineering is responsible for managing and administering that part of the contract for design, procurement and commissioning of the transportation system and fixed equipment within the tunnel system and on the two terminals. Construction is responsible for managing and administering the contract for the design and construction of all civil engineering and building works, and the installation of fixed equipment. Both organizations cover the whole range of technical and commercial issues ranging from design approval to certification of payment applications. Transportation Systems Engineering and Construction account for three-quarters of the PID staff, which averaged 330 during 1991.

31. Eurotunnel's Commercial Manager has a direct link to the contractor's Commercial Director, and with a small and highly specialized support team he deals with major commercial and contractual issues.

32. The Project Controls Manager provides the top-level link to the contractor's project planning function and produces overall project status and progress reports for internal and external distribution. This group also supports the planning and cost functions within Transportation System Engineering and Construction.

33. The Bank Relations Manager manages the interface between Eurotunnel and the banks, and keeps them informed of the status and progress of the project.

External influences

34. The Fixed Link is a high-profile project, and in addition to unremitting media attention it has also been subject to external influence by banks, the IGC and SA, emergency services and local authorities. The management and co-ordination of responses to these organizations is a major task. The interest of the funding banks, with 75% of the investment, needs no further amplification.

35. Arising from the joint documents produced by the UK and French governments was the requirement for the establishment of an IGC with SA. The terms of reference for the two entities were set out in some detail. The IGC responsibility is '. . . to supervise, in the name of and on behalf of the two governments, all matters . . .', and its functions include

Fig. 4. Eurotunnel's Project Implementation Division

(a) monitoring the construction and operation;
(b) approving proposals made by the Safety Authority;
(c) drawing up, or participating in the preparation of, regulations;
(d) giving advice and making recommendations to the two governments or the concessionaires.

Each government appointed half the members of the IGC, which number 16 at most, including at least two representatives of the Safety Authority. Chairmanship alternates annually between the delegations.

36. The Safety Authority is similarly constituted and its purposes include

(a) giving advice and making proposals to the IGC, at the request of the IGC or its own initiative;
(b) participating in the drawing-up of any regulations;
(c) ensuring that the safety measures and practices comply with the national or iternational laws;
(d) examining reports concerning any incident affecting safety and making such investigations as are necessary.

37. In order to support these terms, the concessionaires must provide the IGC and SA with 'avant project' and with safety reports. These documents must provide a great deal of information on a specific selected list of technical, administrative and financial topics. All responses to these submissions must be complied with since (cumulatively) they will represent the total specification of the Fixed Link for which the IGC, in due course, will be responsible for issuing the necessary operating certificate. All formal submissions are required to be commented on by the MdO.

38. The national emergency services exerted an unusual external influence created by the fact that the two sovereign nations are being connected by the tunnel. For this reason the national emergency services took a special interest in the progress of the tunnel, since they had to be ready to react in a co-ordinated way to any emergency in the tunnel. Hence the contractor was obliged to make arrangements for familiarization of the national agencies with the tunnel, and to arrange and co-ordinate training exercises on and around the sites, while at the

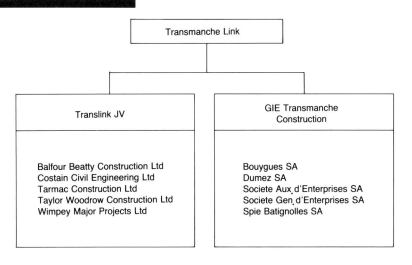

Fig. 5. Transmanche
Link structure

ment has been a continuous process requiring the development of a large degree of co-operation and trust. To assist in that process, the Local Planning Authorities agreed, in a separate Memorandum of Understanding, to acknowledge the overall timetable for the project. Upon the submission of details they were expected to undertake the processing of detailed design and other information in approximately two months.

41. Three of the concerned authorities deal with submissions in the normal manner, referring them to their respective committees. One, however, set up a special Fixed Link Executive Committee which was called to respond to the programme of project submissions appropriately, not necessarily to suit their formal Planning Committee dates.

same time dealing with media speculation as to what was going on.

39. The role of the local authorities was set with the passing of new legislation (the Channel Tunnel Act, 1987, incorporating a Planning Clause and relevant Schedules) which (upon Royal Assent) granted outline planning approval for the proposed works. The Act required, however, that virtually all aspects of the detailed design of the above-ground works and much of the manner in which the construction was carried out also had to be submitted to the Local Planning Authorities for their approval.

40. Within the chosen context of the design and construct project these procedures have implied that Local Planning Authority involve-

Role of the contractor

Formation of TML

42. The Channel Tunnel Group (CTG) was formed in 1984. It consisted of Balfour Beatty, Costain, Tarmac, Taylor Woodrow and Wimpey, together with the National Westminster and Midland banks. At the same time in France a group known as Francemanche SA (FM) came together. Partners were Bouygues SA, Dumez SA, Société Auxiliaire d'Entreprises SA, Société Générale d'Entreprises SA, Spie Batignolles SA, Banque Indosuez, Banque Nationale de Paris and Credit Lyonnaise.

43. The two groups, CTG and FM, came

Fig. 6. Main phases
of project

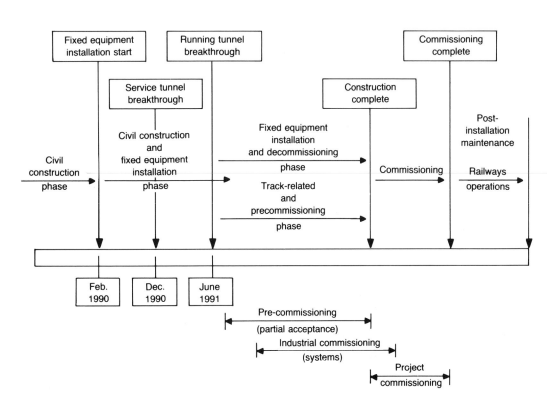

together to put in a proposal to the governments on 31 October 1985 for the twin-bored rail tunnel scheme. It became clear that for the project to be funded privately an independent owner-operator would be required, who in turn would let a single construction contract to the ten contracting companies. Out of this need, Eurotunnel evolved from the original Channel Tunnel Group and Francemanche SA, the ten contracting companies coming together as Transmanche Link.

44.   For legal and fiscal reasons, TML is a transparent (non-trading) company, and so trading is through the two operating companies of Translink JV and Transmanche GIE in Britain and France, respectively (see Fig. 5).

Contract

45.   TML's contract is to design, construct, test and commission the fixed rail link. The link includes twin-bored rail tunnels of 7·6 m internal diameter, and a service tunnel of 4·8 m internal diameter.

46.   The terminals in the UK and France provide a facility for receiving and dispatching the shuttle trains, each 750 m long. The rolling stock and locomotives are also provided under the contract. The contract, which was signed on 13 August 1986, provided for the link to be fully commissioned during 1993. The main phases of the project are shown in Fig. 6.

47.   The contract price is in three parts

(*a*)   target works, including all tunnel work, and this includes the manufacture of all permanent tunnel linings; the target works extend from portal to portal and can be

adjusted by variation orders;

(*b*)   lump sum works, including the construction of the terminals, and also the installation of all fixed mechanical and electrical equipment in both the tunnels and the terminals which can also be adjusted by variation order;

(*c*)   procurement items—all shuttles, locomotives and wagons (the rolling stock) are paid for on a cost-reimbursement basis with a procurement fee paid.

Building up the team—the early management structure

48.   The development of TML has passed though two distinct phases. The first, formed from the award of the concession through to the radical restructuring during 1989, is described here and is illustrated in Fig. 7.

49.   Under the Chief Executive, the directors généreaux established the two trading companies in Calais and Folkestone; these were the Construction groups. Engineering and Transportation groups were formed at the project liason office in Croydon. The teams formed under their respective group directors were responsible for transforming thoughts into reality, and they achieved a high degree of success, without which much of what was to follow could not have been accomplished.

Development of UK construction organization

50.   The UK Construction group originally comprised six sub-projects

(*a*)   tunnel construction;

(*b*)   tunnel support services;

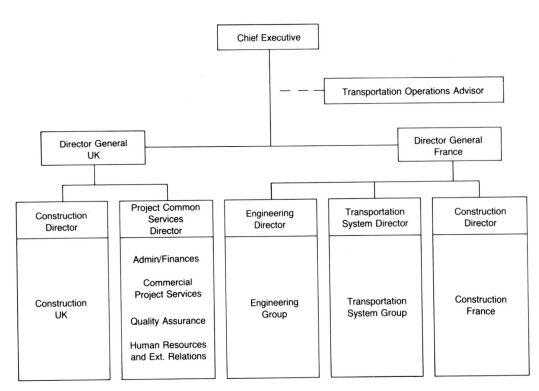

Fig. 7. TML's original management structure

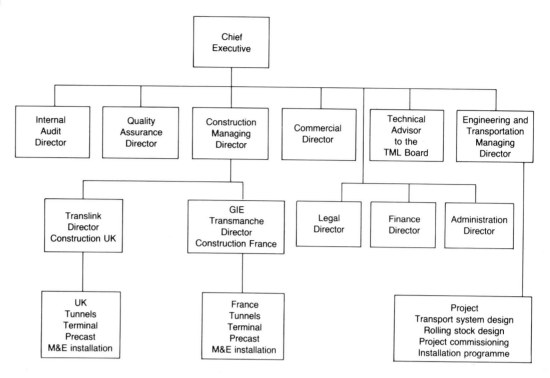

Fig. 8. TML organization chart

(c) precast segment manufacture;
(d) Folkestone terminal;
(e) Ashford works;
(f) M&E installation.

Each operated with considerable autonomy under its Construction Manager, but with a line and function matrix structure linked to the overall group management at the Folkestone headquarters.

51. Under new management, though retaining several key personnel from the original team, the group was restructured as part of the general reorganization of TML during late 1989. The new TML Chief Executive favoured the more pyramidic organization depicted in Fig. 8, while retaining and extending the matrix structure.

52. The UK sub-projects were consolidated into three units

(a) tunnel construction;
(b) terminal construction;
(c) precast manufacture.

Management and administration of mechanical, electrical and track installation subcontracts now form part of the sub-project's scope of responsibility.

53. The French Construction group is similarly structured. Engineering of fixed equipment and rolling stock is the responsibility of the Transport Systems and Engineering group, within which sub-projects are formed for each primary system and discipline. This group was initially responsible for all M&E systems procurement and subcontracts but more recently the responsibility for the basic systems was transferred to the

construction groups. Responsibility for the intelligent systems is retained by the TSE group together with the engineering responsibility for all systems.

Carrying out the work

54. *Design.* The contract defines in its various schedules the basic scope of works. These schedules together with the engineering design requirements and the operating policy of the employer provide the basis for the contractor to develop the design through the prescribed phases of outline, definitive and detail design.

55. As part of the design process the employer is required to ensure that the requirements of the various statutory bodies, national and local, and in particular the Inter-Governmental Commission, are satisfied.

56. The Engineering group established from commencement carried out the development studies to identify optimized solutions and to enable the employer to satisfy its statutory obligations. This process became increasingly complex for the transportation and fixed equipment systems in comparison with that for civil and building works. At the end of 1989 it was necessary to redeploy the Engineering group such that the remaining outline and detailed civil works design became an integrated part of the construction groups and the M&E/transportation design assigned as previously described.

57. For all engineering disciplines much of the design work has been carried out by external consulting organizations working to the direction of TML management and co-ordination teams.

58. *Construction—UK tunnel works*. The detailed aspects of tunnel design and construction are dealt with elsewhere, but it should be recorded that despite early problems, the completion of linear tunnelling was achieved in June 1991 ahead of 1986 programme dates. This tunnelling achievement and the historic breakthrough were recognized by the media, but this gave no indication of the major management task which TML faced in order to achieve this. Some of the facets of this task are worth mentioning.

59. Six TBMs were used, three on the land tunnels and three on the marine tunnels. Early problems with the TBMs were overcome, with the result that the UK service tunnel machine achieved the longest project drive of 22·3 km and the running tunnel machines achieved unprecedented progress rates.

60. A new facility was established at the Isle of Grain to produce the precast concrete segment tunnel linings. The management and control of this facility was crucial in meeting the requirement that tunnel progress must not be constrained by late delivery of segments. A total of 1·6 m tonnes of segments were produced.

61. Logistical support for tunnelling operations of this magnitude was complex and crucial. The system had to support the drives by moving men, materials and tunnel spoil, a task which became more complex as the tunnel drives progressed.

62. The recruitment, training and management of a direct labour force, excluding staff and subcontracts, reached 4100 at the peak of operations on the UK side. Procurement of equipment and materials cost £157 m. There was subcontract placement and management for 114 subcontracts with a value of £284 m.

63. Maintenance support was supplied by means of large plant workshops: one for general plant, one for underground locomotives and one for rolling stock. In total there was a direct labour force of 670 men engaged in mechanical/support activities.

64. The organization, management and control of all the above in a timely and effective manner was crucial to the tunnel achievement.

65. *Construction—terminal*. Achievements at the Folkestone terminal, now plainly visible, have gone largely unsung by the media and industry, as tunnelling work attracted the limelight. This section of the works has proceeded on its 140 hectare site, to become an impressive feature of the project.

Management functions

66. *Construction management*. In early 1990 it became possible for a much reduced group management team to relocate at Shakespear Cliff, as the project liaison office transferred to Folkestone. The reduction resulted from a process whereby individual group managers and support staff were targeted towards the key tasks at the sub-project sites, providing direction and assistance where it was most needed.

67. This more cohesive group organization paved the way for closer inter-group activities as M&E installation came into play, together with cross-border operations between the French and British construction teams. Management policy is directed at weekly meetings of the Executive Committee, attended by the Chairman and Chief Executive with the group and services directors, at which all matters affecting the project are discussed. Policy dissemination is through group and sub-project management meetings. Formal meetings with the employer at executive level are biweekly and monthly at group and sub-project level, supported by a series of task-oriented meetings.

68. *Commercial support*. The commercial department is responsible at executive level for contract administration policy, funding application and legal matters. At group level are added procurement, sub-contract administration and estimating functions. Management finance and cost management are separately managed but closely allied. Group commercial directors are either directly or functionally responsible, depending on size and scope, for the commercial teams within the sub-project management structures.

69. *Human resources*. On the UK side of the project the Human Resources Department's huge recruitment task culminated in approximately 8400 people employed at the end of 1990. Of these, 1800 were in staff position, many of them being seconded from the five Joint Venture Partners. In line with TML's policy to recruit locally wherever possible, 49% of all employees came from Kent (Fig. 9).

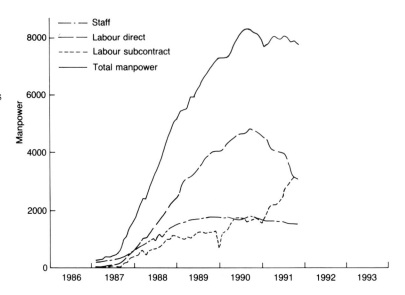

Fig. 9. UK manpower

Planning and programming

70. The bi-national nature and size of the project led naturally to a hierarchical planning structure. M&E design and the rolling stock were treated as project-wide, whereas in the UK and France the project was divided into sub-projects: tunnels, terminals and precast facilities.

71. Bi-national, broad, strategic planning was known as level 1, and national strategic planning linking the sub-projects as level 2.

Fig. 10. MdO organization

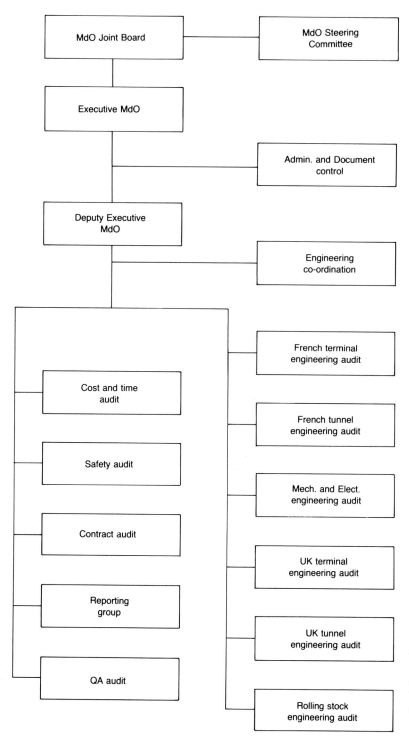

There were specific level 2 links between the French and UK tunnel sub-projects. Construction planning was developed in detail at levels 3 and 4, the latter embracing four week, day by day, rolling programmes and short-term detailed co-ordination programmes. For the tunnelling work the time chainage format was used at all levels as the main strategic planning tool, with much interaction between levels.

72. The whole project was networked at level 2 to provide a firm reporting and monitoring base. However, with the fast-track nature of the project, even this base had to be revised at six to nine month intervals. In 1991 the project-wide level 2 network contained 5000 activities, and the UK tunnel sub-project level 3 network contained over 6000 activities. At all levels computer graphics were widely used to give management a clear view of the programme strategies and progress achieved.

Environmental control

73. Environmental issues were fully integrated in the design, planning and construction of the works. An in-house team of three environmental scientists defined and controlled the work of a dozen specialist sub-contractors. Their task has been to minimize the environmental effects of the works, within the project programme and budget.

Quality assurance and control

74. The principles of the UK TML quality system are based on BS 5750, and are set out in a joint declaration between TML/ET entitled 'Management for quality', framed in 1990. This defines quality in terms of a nine-point programme which places the basic responsibility for quality work on individual team members identifies the need for a planned approach to quality achievement, and emphasizes the importance of quality awareness in the management process and working environment.

Safety

75. A safety culture has been developed on the project which embraces a philosophy of proactive safety management. A 13-strong department of safety advisers plus a site medical team of 20 all report to the Health and Safety Director, deploying a wide range of skills in support of management. Emphasis is placed on accident prevention and safety awareness through training and management audit. This programme, intensified during 1990, is now producing results which attract high commendation and the close attention of the UK construction industry at large.

Role of the Maître d'Oeuvre

Organization and role

76. The background and early development

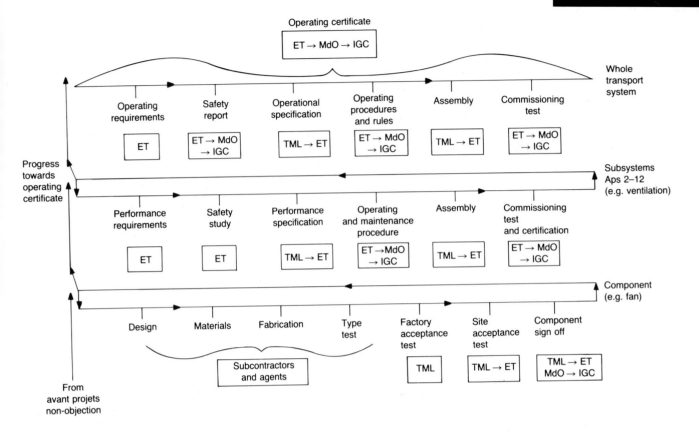

Fig. 11. Conceptual model of operating certificate

of the role of the MdO has already been presented to the Société des Ingénieurs et Scientifique de France and the Institution of Civil Engineers at an international conference held on 22 Sept. 1989—The role of the Maître d'Oeuvre, H. Grimon and P. Middleton. That paper describes the classic MdO management system as used in France and the sequence of events leading up to the role as described at that time. The key elements of the MdO's role may be summarized as follows.

(a) All submissions made by Eurotunnel to the IGC/SA are reviewed by the MdO with particular reference to the requirements of the concession. In this review emphasis is placed upon technical standards, safety and quality assurance.

(b) The MdO prepares a quarterly report for the banks giving an overall factual progress report together with subjective views on any outstanding problems.

(c) The MdO also makes independent technical analysis of engineering problems for discussion with Eurotunnel, and because of the early history of the work is involved in independent assessments of contractual issues.

77. The current MdO organization is shown in Fig. 10. There are six main functions, each interrelated with the others to provide the overall view of the project required by the audit function.

78. The project engineers group consists of one engineer for each of the main sub-projects, e.g. terminals, tunnels, etc. They are responsible for reviewing all technical submissions to the IGC/SA, liaison with the site engineers, and the production of the technical aspects of the quarterly report to the banks.

79. The safety group contains experienced safety engineers responsible for reviewing the safety aspects of the project, and particularly to assess the safety case.

80. The cost and time group reviews the cost aspects of the construction contract and prepares a regular report on anticipated cost at completion. Programmes are analysed and progress is monitored; contractual aspects of the construction contract are assessed.

81. The construction engineers based at the sites review the construction, and will later support the commissioning effort, which is dealt with as a separate aspect of the work. The QA group is responsible on behalf of IGC/SA and ET to ensure that QA standards are maintained throughout the project.

32. The MdO is a joint venture of two major multi-disciplinary consultancies, W. S. Atkins in the UK and Setec in France, supported by Halcrow and Tractabel. The MdO team consists of consultants from the UK and France and Belgium, in roughly equal numbers; the total staff is approximately 74. The organization is designed to match that of ET and TML.

The operating certificate

83. The governments require that an oper-

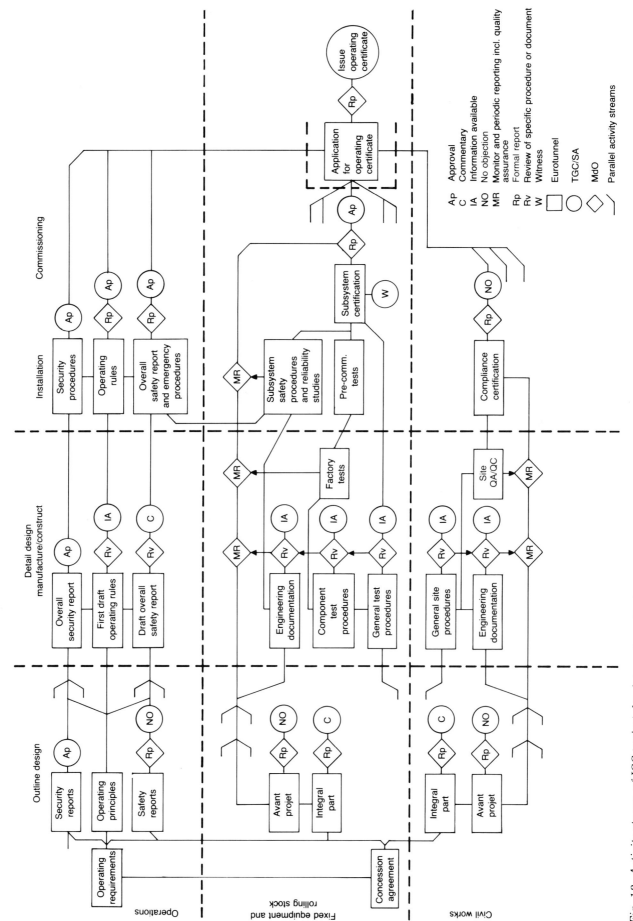

Fig. 12. Activity stream of IGC project development monitoring

ating certificate be issued on their behalf before the overall transportation system can be put into commercial operation and before the public can be admitted. The work of the MdO is designed to lead to establishing sufficient technical confidence in all aspects of the project to enable it to recommend to the IGC/SA that such an operating certificate be issued.

84.   The conditions for achieving this certificate are that

(a)   the project has been built in accordance with the design presented to the IGC/SA and to which they have no objection;
(b)   the appropriate national and international standards relevant to each part of the works have been met;
(c)   the individual items of equipment have achieved their specified performance and safety standards;
(d)   the subsystems operate at their specified performance and safety criteria;
(e)   the transportation system as a whole meets the requirements of the system design and conforms to a developed and agreed safety case;
(f)   the statutory standards of the relevant country have been met, and where both countries are affected that a joint standard is met;
(g)   the appropriate project records, operating rules maintenance manuals, etc., are available.

85.   The SA is responsible for a general overview of these factors and the MdO is responsible to witness each of the above aspects in appropriate detail to ensure that the requirements are met. This task is illustrated in Figs 11 and 12.

86.   Figure 11 shows the sequence of events from individual items of equipment through subsystem tests to the overall system tests. The diagram is very simplified, and in real life will be expanded into a considerable amount for documentation.

87.   Figure 12 shows a similar sequence of events but divides the project into civil works, M&E equipment and rolling stock and documentation. It is clear that signalling and other critical aspects require a much greater level of supervision than, say road and bridges in the terminals, and this diagram illustrates the level of effort required from the MdO to each of the three aspects.

Conclusion

88.   It is hoped that this Paper has given some insight into the way in which the management of the project has evolved since the signing of the contract in Aug. 1986. In assessing the achievement on the management of the project, it is worth restating the initial stages of development of the organizations involved.

89.   TML formed two operating companies, one in the UK and the other in France. In each case the operating company comprised five contractors, with the major task of integrating staff from member companies together with specialists brought in from other sources into an integrated project team with access to the appropriate project control systems.

90.   At the time of contract the client was virtually non-existent and therefore had to use the MdO for support and interface with TML. The formation of the Eurotunnel Project Implementation Division enabled the MdO to return to its independent role. The PID was formed by a mix of personnel from W. S. Atkins, Setec and Bechtel and Eurotunnel. As with the contractor there was the task of integrating these staff into an effective project team with appropriate project monitoring systems available to them.

91.   The creation of the Eurotunnel Project Implementation Division allowed the MdO to revert to its independent role, advising the Inter-Governmental Commission, the Safety Authority and the banks. In addition, the MdO undertakes independent technical analysis of engineering problems and assessments of contractor issues.

92.   It is of great credit to the senior management of the three organisations, supported by the professionalism of all the staff involved, that effective management of the project has been achieved. Inevitably it took some time to achieve this and the TML reorganization in 1989 made a significant contribution to this. From the start clear identification of counterparts in each organization has avoided the conflicts and confusion which often occur with less structured approaches.

93.   A project progresses through a number of distinct phases, and organizational changes need to be made to ensure that the respective teams and their interfaces are relevant to the main area of work being carried out at the time. The project is now moving into the testing and commissioning stage, requiring further progressive changes in organizational structure.

Proc. Instn Civ. Engrs, Civ. Engng, Channel Tunnel, Part 1: Tunnels 1992, 18–42

Paper 9963

Tunnel design and construction

G. S. Crighton, DSc(Hon), BSc, FICE, A. R. Biggart, BSc, FEng, FICE, and E. H. Norie, MA, FICE

G. S. Crighton, Engineering Director, UK Construction Transmanche-Link JV

A. R. Biggart, Operations Director, UK Construction Transmanche-Link JV

E. H. Norie, Director, Mott MacDonald Ltd

■ **The work described in this Paper has been contemplated by engineers for nearly 200 years. The scope of work carried out under the target cost part of the contract covers the design, construction, testing, commissioning and maintenance of the tunnel works. This includes all necessary temporary works, the provision of linings, the disposal of large quantities of spoil, and the mobilization of vast quantities of construction plant. Under the lump sum section of the contract the tunnels have been converted into a railway system by the installation of the fixed equipment. This covers the track, signalling system, overhead catenary, power, ventilation and drainage system, and the communication and control network. From the start it was recognized that the main challenge of the project was in solving the complex logistics needed to support the high-speed tunnelling operation. The other major challenge has been the fast-track nature of this design and construction contract. This Paper provides an overview of the design and construction of the UK tunnel works, and it attempts to cover the above issues.**

Concept and scope of tunnelling works

Present scheme
The present scheme was chosen to provide a Fixed Link across the Channel and was the outcome of the competition organized by the British and French governments in 1985. Three main contending proposals had been put forward by different sponsoring groups, which embodied alternative approaches to the creation of a Fixed Link for road or rail track. The concession to build and operate the Fixed Link was awarded to the Channel Tunnel Group (CTG) (subsequently to become Eurotunnel (ET) and Transmanche-Link JV) in March 1986. A contract for the design, construction, equipping and commissioning of the Fixed Link was awarded to Transmanche Link JV (TML) later in that year.

2. The terms of its contract with Eurotunnel required TML to have the Fixed Link operational by mid-1993, giving a very challenging programme for development, design, construction and commissioning of the facility. This programme influenced every aspect of the project and created many of the key management tasks.

3. The scheme is based on rail transportation, and it links the rail networks of the UK to those of continental Europe; it will carry both through-trains and special shuttle trains, the latter comprising 'ferry' stock to take drive-on road vehicles. The Fixed Link runs between terminals at Folkestone in England, and Coquelles, near Calais, in France, and lies in a tunnel throughout its length between the seaward ends of the two terminals, a distance of 51 km. The length of tunnel under the sea between the two coastlines is 38 km.

4. A similar scheme had been promoted by the two governments in 1972. The first phase of construction of that scheme was started in 1973, involving on the UK side the development and assessing of working sites at cliff-top and low level at Shakespeare Cliff, the sinking of an adit from the Lower Site to tunnel level, and the driving of a length of service tunnel using a tunnel boring machine. Although this work was abandoned (for political reasons) in early 1975, useful work had been carried out, and information was obtained which facilitated the rapid development of the present scheme. In particular, data on the behaviour of Chalk Marl during tunnelling, the build-up of loads on tunnel linings, and the behaviour of precast concrete lining segments under load were of direct value.

Main elements of scheme
5. *Main tunnels* The scheme comprises two running tunnels, each carrying a single rail track, and a separate service tunnel lying midway between the running tunnels and connected to them by cross-passages. Two separate running tunnels, rather than a single large-diameter tunnel carrying twin tracks, were chosen to minimize construction risks and to provide greater safety of operation and maintenance. The tunnel diameter and spacing were finalized as a result of development and optimization studies, taking account of final function and operation, construction speed and cost, and particularly bearing in mind the servicing during construction of the exceptionally long drives from the coastal working sites.

6. The service tunnel has the primary function of providing access to and from the running tunnels throughout their length in both normal and emergency conditions, and it allows the tunnels to be evacuated, following any emergency halting of trains, within 90 minutes. The service tunnel carries its own guided transportation system. In addition, it accommodates

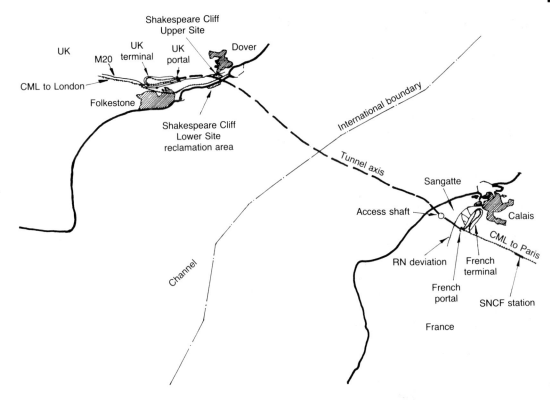

Fig. 1. Layout

water and pumping mains and a drainage channel, and acts as the fresh air supply duct to the tunnel complex in normal operation. It is designed to be kept at positive pressure in relation to the running tunnels in all conditions to prevent smoke ingress.

7. The service tunnel was designed to act during construction as a pilot tunnel, allowing ground conditions to be investigated more thoroughly in advance of the running tunnel drives. It was also used for ground injection to improve running tunnel driving conditions where this was required near the UK coast.

8. The running tunnels each carry, in addition to the rail track and overhead line equipment, power supplies, secondary drainage, cooling pipes, walkways and auxiliary services. The running tunnels act as the main air ducts for the supplementary ventilation system, used in particular to control the direction of smoke emanating from a fire.

9. *Cross-passages and equipment rooms*   The cross-passages connecting the running and service tunnels are at 375 m centres, such that three passages will be directly adjacent to a halted shuttle train. The cross-passages incorporate fire-proof evacuation doors which may be opened and closed in all operational conditions.

10. Equipment rooms, accommodating transformers, switchgear and signalling equipment, are located between the service and running tunnels, and are accessed from the service tunnel.

11. *Pressure relief ducts*   Differential air pressures and aerodynamic resistance build up rapidly on a train travelling at speed in a very long tunnel. The shuttle stock in particular gives a high blockage ratio (ratio of train to tunnel cross-section); without pressure relief the power required to drive a train at operational speed would greatly increase.

12. The pressure relief system adopted comprises 2 m diameter ducts at 250 m centres connecting the two running tunnels, crossing above the service tunnel. Their function was modelled extensively during design to achieve a balance between aerodynamic efficiency and sufficiently low wind velocity in the ducts to avoid undue buffeting. The ducts are fitted with fully closable dampers to prevent the passage of smoke from one tunnel to the other in case of fire.

13. *Pumping stations*   Five pumping stations serve the whole of the underground works: three under the sea, one at Shakespeare Cliff and one at Sangatte.

14. *Crossovers*   For operational and maintenance requirements, four double crossovers are incorporated between the two terminals to allow trains to cross from one running track to the other; two crossovers lie close to the two terminals; the other two lie under the sea and divide the tunnel length into three approximately equal sections. The undersea crossovers take the form of very large chambers at which point the two running tunnels come close together, the service tunnel being diverted

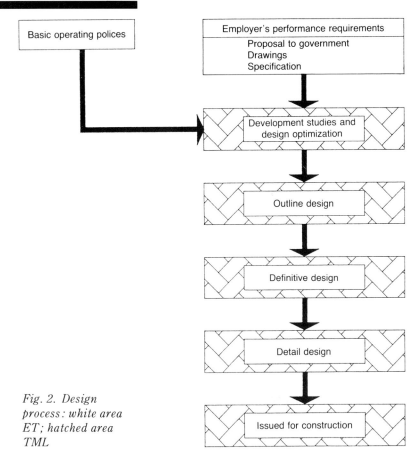

Basic operating polices

Employer's performance requirements
Proposal to government
Drawings
Specification

Development studies and design optimization

Outline design

Definitive design

Detail design

Issued for construction

Fig. 2. Design process: white area ET; hatched area TML

imately 500 m long. The western portal forms the permanent UK portal to the Channel Tunnel. The tunnel layout is shown in Fig. 1.

Development, organization and programme

Design process

17. The main phases of the design process are shown in Fig. 2. Each phase was submitted by TML for approval by Eurotunnel.

Design development

18. It was recognized that the submission made to the governments in October 1985 in response to the invitation earlier in the year represented a viable solution to the tunnel. However, TML was required to reappraise the submission to ensure that it represented a balanced solution in terms of safety, comfort, speed, operating cost and capital cost, with a view to maximizing the commercial viability without increasing prices, costs or the time schedules. This reappraisal was carried out in the Development Study Phase.

Development study phase

19. The aim of this phase was to define the design criteria and optimize the design of the works. The main development studies undertaken were

(a) train speed and traction power
(b) ventilation and cooling in tunnel
(c) running tunnel diameter, aerodynamic effects and piston relief ducts
(d) wagon design, including wheel design and passenger comfort
(e) tunnel gradients (items (a) to (d) were considered to be inter-related)
(f) tunnel lining design
(g) tunnel earthing and bonding
(h) system control and communications
(i) track form
(j) overhead current collection
(k) number, location and construction of tunnel crossovers
(l) shuttle lengths, weight and composition, including numbers of rolling stock
(m) service tunnel transport systems for both normal in-service and emergency access/ evacuation
(n) UK terminal layout including 'free exit' and integration with BR rail and the road network
(o) French terminal layout and integration with French National rail and the road network
(p) maintenance layout
(q) geotechnical investigations for terminals and tunnels
(r) safety.

locally to one side. Each chamber contains a diamond crossover designed for 60 km/h running. The chambers are equipped with fire doors to close the full length of the opening between the running tracks in normal operation.

15. *Tunnel drives* The seaward and landward drives for all three tunnels on the UK side were carried out from Shakespeare Cliff, with all spoil disposal being to lagoons behind a new seawall at the Lower Shakespeare Site. Access to tunnel level for construction traffic was by means of a new inclined adit (A2) from the lower site. A shaft, for man access, was sunk to tunnel level from the Upper Site. A network of marshalling tunnels was constructed at tunnel level to service the main tunnel drives.

16. The UK seaward drives are between 18 and 22 km long from Shakespeare Cliff to the mid-Channel junction with the French drives. The UK landward drives are 8 km long to Sugarloaf Hill at the eastern side of Holywell. Holywell is a re-entrant valley 500 m wide between Sugarloaf and Castle Hill, and it has a high ecological and environmental value. It is traversed in cut-and-cover, with final restoration of the land surface to its original condition. Castle Hill is an outlier of the North Downs at the eastern end of the Folkestone Terminal. It is traversed by bored tunnels approximately

The overall study plan and the specific study plans were established and agreed with Eurotunnel, as were the final development study reports before their incorporation in the design process.

20.   A design development process is illustrated graphically in Fig. 3 for the crossover, and it is typical of the process undertaken for all design work on the project.

Outline design phase

21.   The outline design built on the design criteria and optimization obtained from the development study. It also served as the basis for submissions to the Inter-Governmental Commission for approval in accordance with the terms of the Concession Agreement.

Definitive design phase

22.   The definitive design developed from the outline design and took account of comments made by ET. Where design-and-construct packages were awarded, the definitive design documentation was used as the basis for the technical content of the tender documentation.

Detail design phase

23.   The detail design resulted in the production of the detailed documentation (drawings, specifications, criteria, etc.) necessary for the construction or fabrication of the finished works. In all, it is estimated that project-wide over 100 000 design documents will have been produced by the end of the design phase. A specially developed technical documentation register available at any authorized PC or system terminal allowed instant access to all information concerning design documents.

24.   The fast-track nature of the project required functional studies to be carried out at the same time as the civil structures design, and the civil design to be carried out as construction proceeded. The design-and-construct type of contract awarded to the contractor, holding one party responsible for all these activities, permitted the extremely tight design and construction schedule to be maintained.

Site investigations and general geology

25.   The geological plan and section is shown in Fig. 4. The geotechnical information shown is based on the results obtained from all geotechnical investigations which have been carried out in the Channel since 1875.

26.   TML undertook two further geotechnical and geophysical surveys in 1986–87 and in 1988 along the proposed line of the tunnel. The main objective of the 1986–87 survey was to improve, update and supplement the existing information, particularly in areas where signifi-

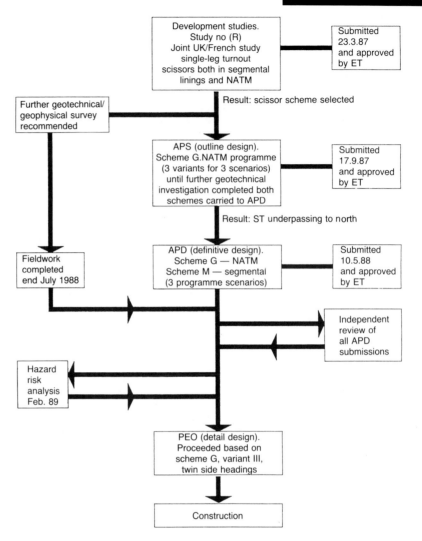

cant changes had been made to the alignment of previous schemes.

27.   Because of the buried infill valley system at the Fosse Dangeard and the major excavation proposed for the crossover, it was decided that more detailed site information was needed in these areas, and further surveys were carried out in 1988.

28.   Both the 1986–87 and 1988 investigations included boreholes and a geophysical survey. Considerable advances in geophysical techniques have occurred in the last 15 years, owing to improved equipment and the application of computer processing of the data. This is the first time that these methods have been applied to a major civil engineering project. The geophysical surveys had two main objectives. Firstly, to determine the shallow geology to 150 m below the seabed with a greater accuracy and precision than had previously been possible; and secondly, to investigate the deeper sub-tunnel geological structure to approximately 800 m below seabed, to provide data for seismic risk evaluation.

29.   Owing to the relatively shallow band of

Fig. 3. UK crossover design development

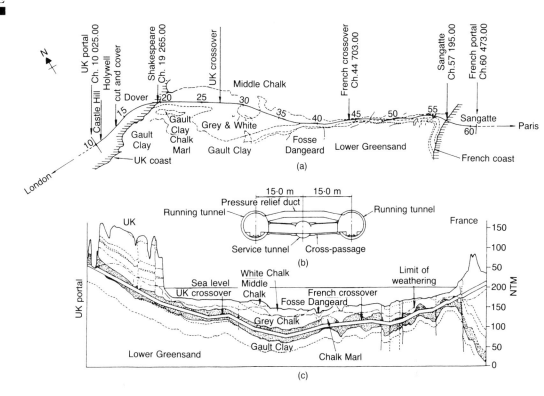

Fig. 4. Layout and geological section; (a) plan; (b) typical section through tunnel; (c) longitudinal section

chalk marl able to accommodate the tunnel, it was important to establish the accuracy of the geotechnical predictions of strata and feature locations. To this end a detailed reliability analysis of the component elements of the geophysical survey work was carried out, including bathymetry, positioning accuracy, signal timing and velocity data, data accuracy weighting, and the quality of processing and data combination procedure.

30. The use of a large North Sea oil rig shown in Fig. 5 to obtain boreholes in the Channel greatly assisted in their accurate positioning. It was fortunate that the borehole investigation coincided with a downturn in North Sea oil exploration, enabling such rigs to become available at competitive rates.

31. The 1986–87 and 1988 surveys covered 19 boreholes in the Channel at an approximate average cost of £½ million each and approximately 1000 kilometres of seismic survey costing approximately £1500 per kilometre.

Materials and equipment ordering

32. Procurement of the target works, tunnelling, element of the project was a major task. The plant holding alone rose to a value of £165 million. The main items of construction plant are shown in Table 1 alongside the main project statistics for civil works and fixed equipment.

33. All procurement for target works was subject to procedures which involved approvals from Eurotunnel. Before the procurement process could be completed many engineering decisions were needed. Although much of the plant consisted of standard off-the-shelf items, there were a number of purpose-built items;

these included the tunnel boring machines, the electric locomotives, the spoil disposal conveyor system, the 20 tonne portal cranes for segment unloading, the precast factory carousels, the 12 tonne portal cranes and the steel moulds.

34. Bearing in mind the fast-track design and construction nature of the project, the basic engineering decisions regarding excavation and ground support (together with spoil disposal and precast concrete segment casting) needed to be taken at a very early date. It says much for the engineers and managers present at the early inception stage that these important decisions turned out, on the whole, to be right. The tunnel boring machines and electric locomotives needed on-site modification, some of which was due to the need to deal with wetter conditions than expected.

Interfaces

35. For this fast-track design and construction contract it was necessary for TML to address a large number of internal and external interfaces. The relationships between the various bodies were constantly changing as the project passed through the stages between initial concept, detailed design, civil construction, fixed equipment installation, commissioning and final operations. In order to co-ordinate the interfaces between internal departments, senior managers on both sides of the channel were appointed to enable the right level of internal communication. Dealing with some of the external interfaces required specialist knowledge, and small dedicated teams were set up for this purpose. These specialist teams

included those dealing with: the Inter-Governmental Commission through Eurotunnel, to whom 37 major system design submissions were made; the Planning Authorities, to whom 360 planning applications were made; and the Environmental bodies, of whom there were 35 with an interest in the works. Other outside agencies were of the sort that could be expected on any large project. However, the transparency of all that was done on the project and the high political and social profile required that even the normal interfaces required more than the usual attention.

36.   The main interfaces for the tunnel sub-project are shown in Fig. 6.

Engineering and construction organization

37.   *Pre-construction phase*   The general framework of the project provided for an operational division of the project into the following areas: tunnels; precast concrete; terminals; mechanical and electrical; and operations. Except for operations, the engineering and construction of each area were managed as separate sub-projects reporting independently to the Engineering Director and the Construc-

Fig. 5. Boreholes in Channel

Fig. 6. Main interfaces

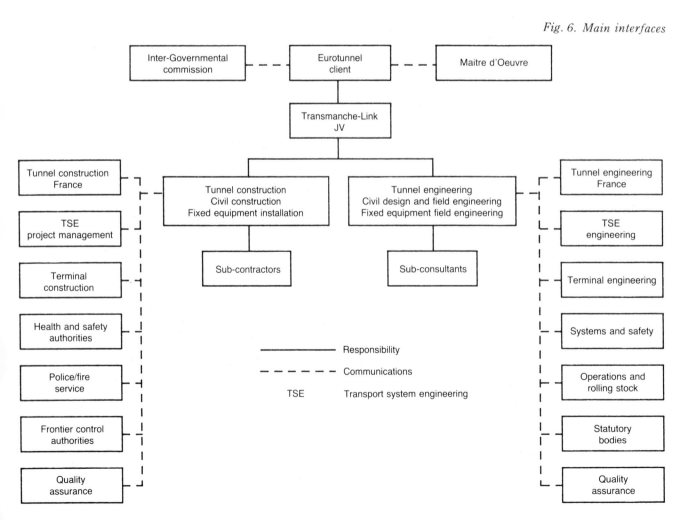

tion Director (see Fig. 7).

38. The tunnel engineering sub-project was responsible for all actions concerning the engineering options, the planning schedules, the costs relating to engineering solutions, mechanical and electrical co-ordination and the design consultants and budgets. The optimization of costs and adherence to the design programme were two important factors of the contract requirements.

39. The tunnel construction sub-project was responsible for procurement, detailed construction planning and programming, and the choice of construction methods based on engineering advice from the tunnel engineering sub-project. Obviously, close co-operation was necessary between the tunnel construction sub-projects (UK and France) and the tunnel engineering sub-projects (UK and France), as well as the holding of regular formal co-ordination meetings at all levels. The sub-project offices were located together so that day-to-day contact was facilitated.

40. *Construction phase* Once construction had started the various tunnel sub-project groups were located in France and the UK. The UK and French tunnel engineering sub-projects then became responsible to the Construction Directors UK and France, respectively (see Fig. 8).

41. General responsibilities for the tunnel engineering and tunnel construction departments is shown on the organization chart (Fig. 8). Procedures were established to cover all aspects of the construction and design works. These procedures detailed quite clearly the responsibilities of the engineering and con-

struction departments. A flow chart for a field change request to the design is shown in Fig. 9 and is typical of the process. In many cases, owing to the extremely tight construction programme, this type of change needed to be actioned instantly, thereby requiring close co-operation between engineering and construction.

Programme

42. The construction programme drawn up in November 1986 (Fig. 10) illustrates the extremely tight schedule for the tunnel works. It is interesting to note, and is a credit to all involved, that despite numerous unforeseen problems along the way the tunnel civil works achieved the November 1986 programme.

Main design and construction features

Shakespeare underground development

43. It was a fundamental construction decision to drive all the UK-bored tunnels from a single work site. This had many advantages, including a near-at-hand spoil disposal area behind the new sea wall. However, the decision also increased the complexity of the logistics, because it was necessary to provide support to five tunnel boring machines (TBMs) at any one time. These required 1000 precast concrete segments every 24 hours, together with other support materials such as track, cable, pipes and ventilation ducts. At peak, over 18 000 m³ (bulked) of excavated spoil per 24 hours arose from the five TBMs. This logistical challenge was handled successfully by a process of separation: personnel entering by the 110 m

Fig. 7. Organization: pre-construction phase

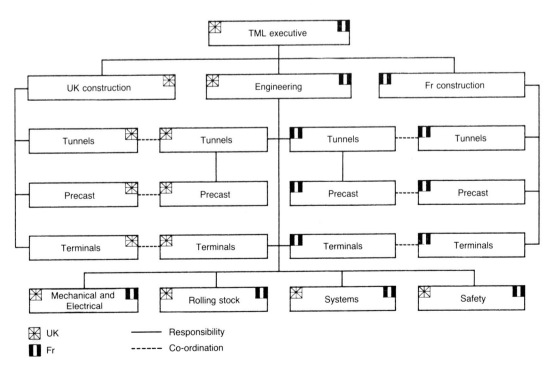

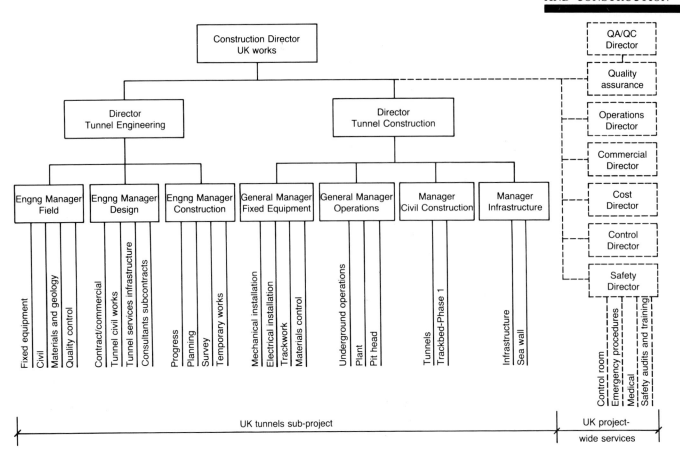

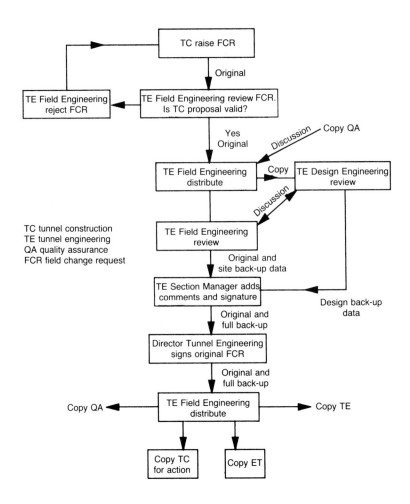

Fig. 8 (above). Organization: construction phase

Fig. 9 (right). Flow chart for FCR

deep shaft; all spoil removal up adit A1 on a 2400 t/h capacity conveyor; and all materials taken in on the five line rack track railway in adit A2. In addition, a highly organized materials-handling system was set up on the Lower Site at Shakespeare Cliff. This received all the precast concrete segments on the British Rail line, which ran along the back of the site, and also bulk materials such as piles and aggregates.

44. The marshalling area, as it was called, had six TBM erection chambers and a complexity of conveyor and ventilation tunnels, nearly all formed in the New Austrian Tunnelling Method (NATM) (Fig. 12). The period of construction was 19 months.

45. The main tunnels in the marshalling area were 500 m long between the marine and land TBM erection chambers. They were excavated by Voest Alpine AMT 70s and Demag roadheaders using a heading and bench approach, which was necessary because of the 9 m height of these drives. The new adit A2 was excavated using the Demag Roadheader, which unlike the Voest Alpine did not have a gathering facility at the front end. It was rightly decided, in advance, that gathering of

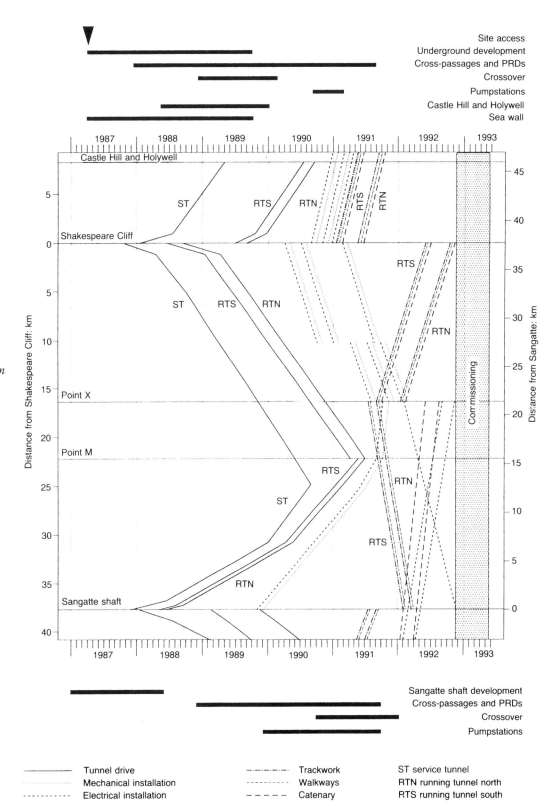

Fig. 10. Construction programme, November 1986

the material, which was constantly wet from the water that ran to the face from the upper wet areas, would have to be done by a front end loader.

46.   All the NATM tunnels have been given a final in situ concrete lining. The excavation quantity for the marshalling area was 160 000 m³, the shotcrete area was 40 000 m², and 8000 m³ of other concrete was placed. Once the secondary lining has been placed, the total concrete quantity in this area will be 21 000 m³.

47.   The programme for the Channel Tunnel project demanded the fastest possible construction techniques to ensure that the TBM drives commenced on time. It was this programme requirement that led to the adoption of the NATM, with the secondary lining construction delayed until completion of the TBM drives and the removal of associated construction plant and equipment.

48.   The Shakespeare tunnels lie predominantly in the Chalk Marl found to be generally dry and favourable for the NATM. The geotechnical parameters adopted before construction were found to be conservative, following observation of the constructed primary linings. This led to more economic and flexible designs for other NATM tunnels at Shakespeare. Indeed the success of the NATM at Shakespeare was a major factor in its adoption for the Castle Hill tunnels, and, more significantly, for the UK undersea crossover.

49.   A particular feature of the NATM is that the design construction method, construction sequence and construction plant are closely interdependent. Successful implementation depends on continuous integration of these separate elements. Engineering personnel monitored construction work continuously to confirm the design assumptions and to provide a rapid on-site response to varying construction demands.

50.   The speed and facility of the NATM is particularly well-suited to excavations of complex geometry such as the marshalling area. The area contains multiple junctions, small and large tunnels, a shaft, and various curved tunnel sections. The choice of an NATM design for this complex layout is dramatically confirmed when a comparison is drawn between the construction time for a tunnel and adit junction constructed by traditional tunnelling methods in 1974, and the construction time achieved for an NATM junction under the present contract. The former was three months, the latter three weeks.

Marine works

51.   All spoil arising from the tunnelling operation was placed and compacted in a dry state behind a 1700 m long sea wall constructed using three unifloat jack-up barges. The 4·25 × 10⁶ m³ of spoil formed a reclaimed area of over 30 hectares which was used during con-

Fig. 11. Shakespeare Cliff

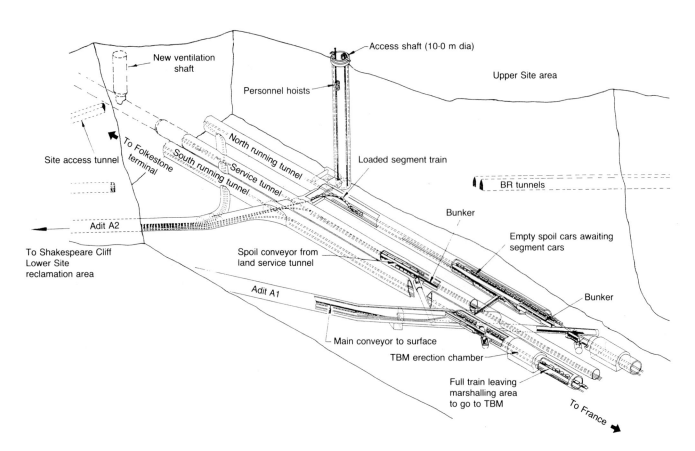

27

struction for material storage and handling and for plant maintenance. During operations it will be used for permanent cooling, ventilation, drainage, water supply and stand-by power facilities, as well as forming an amenity area for the local population. Fig. 13 shows the almost completed reclamation at Lower Shakespeare.

52. In contrast, spoil disposal on the French side was carried out by converting tunnel spoil into a 50% slurry and pumping the slurry 3 km to the Fond Pignon dam above the working site.

53. The design chosen is advantageous in terms of simplicity, method, programme and cost. It consists of a mass concrete in situ wall placed in three lifts within 10 m long Larssen 6 sheet-piled cells. The temporary crosswalls were similar but infilled with gravel.

54. An extensive set of studies was undertaken, examining wave, tidal, hydraulic and current conditions, beach processes, ground and spoil characteristics, and environmental effects. The site is an area of special environmental, nature conservation and scientific interest.

55. The permanent sea wall is designed as a short-term breakwater and a long-term spoil-retaining structure. Overturning and sliding resistance are provided by the weight of concrete and the sheet-piled diaphragms. The wall is founded on sound Lower Chalk of high bearing capacity. The mass concrete is a grade 35 PFA/OPC mix, with favourable heat evolu-

tion and strength-gain characteristics, giving the durability necessary for the exposure.

56. The seaward spoil slope is protected from erosion by concrete paving and energy-dissipating upstand units. Filtered drainage is provided for wave overtopping, rainwater and groundwater, and scour protection is placed at the toe of the seawall. At the eastern end of the reclamation a new armoured slope prevents increased erosion of Shakespeare Cliff by wave reflection from the reclamation.

Castle Hill, Holywell cut and cover

57. At the Folkestone terminal end of the 8 km land drives the tunnels pass through the low-lying area of Holywell and then through Castle Hill. This whole area of the North Downs escarpment contains a series of ancient land-slips. The layout of this area is shown in Fig. 14.

58. Three alternative solutions for the tunnels through Castle Hill were considered. Two utilized segmental linings and one an in situ concrete lining. The use of the underland TBMs with segmental linings, a conventional road header and shield with segmental linings, and the NATM with primary shotcrete support and a secondary in situ concrete lining, were studied in detail during the design phase.

59. Assessment was also made of the tunnelling conditions which would be encountered within the landslip at Castle Hill, and a trial shaft and heading were excavated into the land-

Fig. 12. TBM erection chamber in NATM

slip using the various alternative solutions for ground support. The slip was stabilized in advance by toe weighting. On completion of these trials, the NATM solution was adopted, because it gave greater programme flexibility, indicated a construction cost saving, and also because this lining system incorporated a drainage layer which has the potential to contribute to the stabilization measures for Castle Hill by reducing the pore water pressure of the water table. Secondary concrete linings with waterproof membranes were provided to the NATM tunnel drives using conventional invert and arch pours.

60. The vertical alignment through Holywell dictated that a reinforced concrete box structure, half-buried in the Gault Clay, with fill placed over the upper half, was the most economical solution. The cut-and-cover tunnels are potentially prone to significant heave and differential settlement at different stages of excavation, construction and backfilling. The design was complicated by the incorporation of a single leg crossover between the running tunnels at this point, as well as by the diversion of the service tunnel below the southern running tunnel. It was decided to design the complete tunnel complex for monolithic behaviour, with no permanent movement joints; this allowed deformations due to ground movement to be kept within tight alignment criteria, and did not require significant additional reinforcing steel; construction was facilitated by the

Fig. 13 (right). Lower Shakespeare

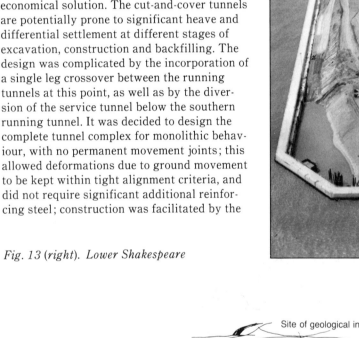

Fig. 14. Castle Hill, Holywell

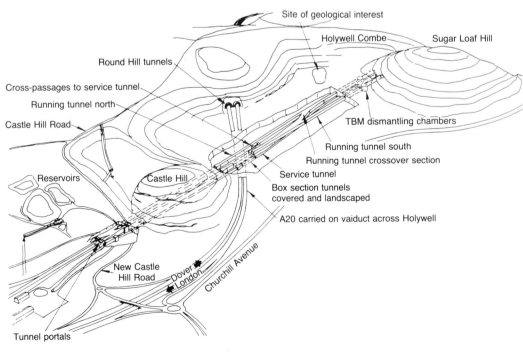

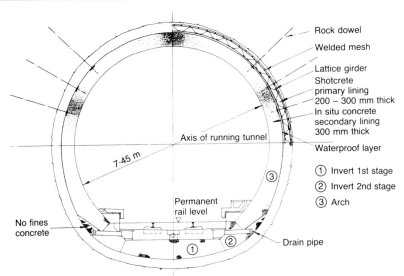

Rock dowel

Welded mesh

Lattice girder

Shotcrete primary lining 200 – 300 mm thick

In situ concrete secondary lining 300 mm thick

Axis of running tunnel

7·45 m

Waterproof layer

① Invert 1st stage
② Invert 2nd stage
③ Arch

Permanent rail level

No fines concrete

Drain pipe

Fig. 15. Castle Hill, running tunnel

elimination of movement joints.

61. At the west end of the tunnels, the drives enter the geologically unstable section of Castle Hill affected by an ancient landslip. Here the excavated cross-section was increased and the secondary lining provided with reinforcement and movement joints as a precaution against distortion. Studies were undertaken to investigate how best to overcome the induced movement in the hillside. Massive imported toe weighting, together with drainage adits, stabilized these movements by the time the secondary linings were constructed.

62. These tunnels through Castle Hill were excavated using a Voest Alpine AMT 70 which discharged spoil straight into dump trucks. The general method of construction is shown in Fig. 15. The tunnel was driven in a mixed face of Gault Clay and Glauconitic Marl with Chalk Marl in the roof; this gave rise to some instability in the working face. This was dealt with by the use of forward-driven spiles in conjunction with lattice girder arches set at a forward rake of 2 m at the top. Spacing of arches was varied to suit ground stability. The running tunnels were driven nearly full face and the invert closed within 12 m of the face. Shotcrete was applied to the sides and face of the excavated tunnel in order to hold the face. The settlement was generally 20 mm and approximately 40 mm at the two portals.

Bored tunnels

63. Six TBMs were employed to drive the UK tunnels a total distance of 84 km. The TBMs were all open-faced machines of the same generic type, with a front excavating section and a rear gripper unit which acts as a temporary anchor point while the front excavating section is thrust forward in 1·5 m increments. Excavation and lining erection are concurrent, so that the machine cycle time is dictated by either the combination of excavation and regripping or by lining erection. Generally lining erection was the criterion of progress. Fig. 16 shows one of the marine running tunnel TBMs.

64. The two service tunnel machines at 5·38 m and 5·76 m outside diameter were both supplied by James Howden. The two marine running tunnel machines at 8·36 m outside diameter were supplied by Robbins Markham

Fig. 16. Marine running tunnel TBM

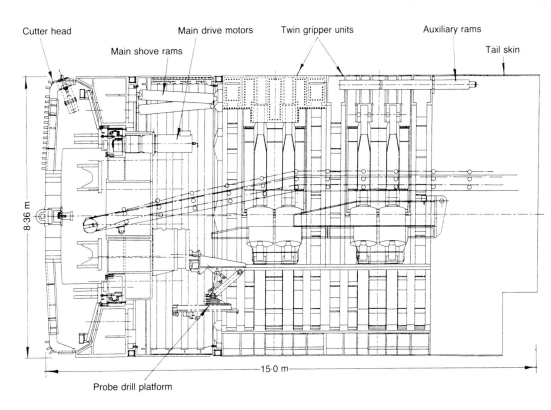

Cutter head

Main shove rams

Main drive motors

Twin gripper units

Auxiliary rams

Tail skin

8·36 m

15·0 m

Probe drill platform

Joint Venture. The land running tunnel machines at 8·72 m outside diameter were supplied by Howdens. The combined weight of the six machines was 6500 tonnes and the total value including necessary spares was £60 million. The longest of the six drives was the marine service tunnel, at 22·26 km. The two marine running tunnels were 18 and 19 km long, and all the land tunnels were 8·0 km long.

65.   The back-up equipment on sledges behind the TBMs was typically 200 m long and contained segment handling and erection equipment and muck disposal conveyors, together with electro hydraulic power packs, ventilation equipment, drainage and water supply and a complex electric power supply system. All TBMs were erected and launched from erection chambers in the marshalling area.

66.   The objective of the tunnelling system was to sustain high rates of progress. This led to the adoption of an expanded concrete lining for the majority of the UK drives; this unbolted lining is expanded directly against the exca-

vated ground. Fig. 17 shows a completed running tunnel. Pads on the back of the lining allow the formation of a 20 mm annulus which was filled with cement grout. This enabled the lining, which is not of itself watertight, to prevent the majority of water ingress. Where openings were required, a cast iron concrete hybrid lining was used.

67.   For the majority of the marine tunnel drives trailing fingers, as shown in Fig. 18, were used in conjunction with thin infill plates which formed a flexible hood. This enabled the lining to be erected and partially expanded under the hood in wet blocky conditions.

68.   Using this tunnelling system the service tunnel machines were designed for an 18 minute cycle, and the running tunnel machines for 22·5 minutes. These cycle times were achieved in practice. Tunnel boring was carried out 24 hours a day, 7 days a week, from December 1987 until the last tunnel was finished in June 1991. Record progress rates were achieved, the best week being 428 m in a marine running

Fig. 17. Completed running tunnel

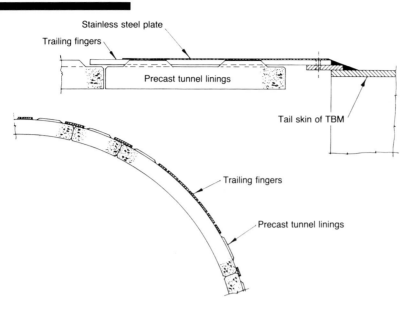

Fig. 18. TBM
trailing fingers

tunnel, and the best 10 week average being
385 m/week. The overall average progress for
all machines was 150 m/week.

69. These rates of progress were achieved
using 4 gangs of up to 47 men, each working 8
hour shifts, on a basis of 6 on and 2 off. In
order to sustain the progress, the gangs
included a maintenance team; the aim was to
keep TBM downtime below 10%. Once the
TBMs had gone through their learning curve
this was generally achieved. Duplication of the
majority of mechanical and electrical
equipment was helpful in this achievement. On
the marine service tunnel a 50 mm probe hole
was drilled forward every 180 m.

70. The French tunnels were driven through
different ground. The tunnels intersected a
number of fault lines and also passed through

the grey chalk layer which lay above the chalk
marl. For this reason a very different type of
TBM, known as the Earth Pressure Balance
Machine, was used. These machines were supplied by Robbins of America and Kawasaki and
Mitsubishi from Japan. Fig. 19 shows the
French marine service tunnel machine. At times
the marine machines, working in closed mode,
operated with a 10 bar pressure in the face.
Owing to the need to be watertight the French
system used a gasketed bolted lining.

*Cross-passages, equipment rooms and piston
ducts*

71. Tunnel connections between the main
running tunnels and the service tunnel were
provided for evacuation cross-passages,
equipment rooms and piston ducts, as shown in
Fig. 20.

72. Spheroidal graphite iron linings were
used for these cross-tunnels and a special
hybrid concrete/cast iron solution was developed and adopted for the openings in the main
tunnels. These hybrid cast iron opening sets
were able to be built at the same rate by the
tunnel boring machines in conjunction with the
cheaper standard pecast concrete linings, and
they proved to be a very economical solution to
openings in the main linings (see Fig. 21).

73. These cross-tunnels were driven by
hand methods using pneumatic spades. With a
compressive strength of 7 N/mm², the Chalk
Marl is hard to dig by hand (London Clay has
strength of about 0·2 N/mm²). For this reason a
number of experiments were carried out which
involved cutting the Chalk Marl with small
track-mounted roadheaders; none of these were
successful. In total, on the UK side there were
245 cross-passages and 106 piston relief ducts,

Fig. 19. French
marine service tunnel
TBM

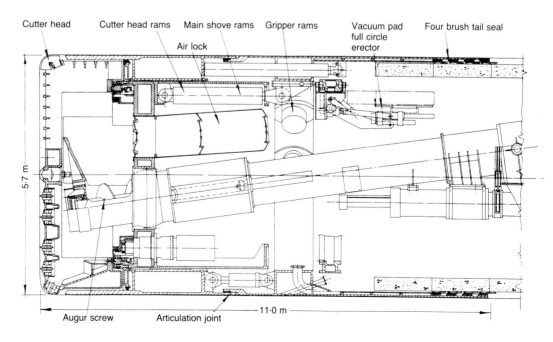

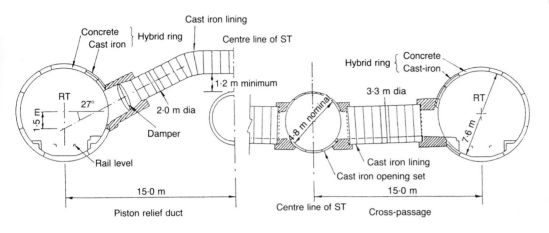

Fig. 20. Connections between tunnels

involving 42 000 m³ of excavation.

74. In order to maintain the contract programme and also provide rooms for temporary pumping stations and electrical sub-stations, it was necessary to construct cross-passages within 2 km of the tunnel face. This meant that the normal double-track temporary railway needed to be reduced to a single track by using a possession on one track; these possessions had a marked effect on the logistical back-up to the TBMs.

75. In situ concrete placed in a single pour was used to form the junction between the bored tunnels and cross-passages.

Pump stations

76. Five main pumping stations and sumps have been provided. Three of these occur undersea and two at each shore, intercepting water from the respective land tunnels and preventing it from entering the marine tunnels. Two undersea and one land pump station are included in the UK works. As well as collecting and pumping out seepage water, these pump stations also impound water that may be caused by leakage of water service pipes or waste water due to fire fighting. In addition, they impound and dispose of any accidental discharge of chemical liquids which may enter the system.

77. Pumping stations are symmetrical around the service tunnel, thus allowing operational duplication between the two sides. This is an important safety and maintenance feature.

78. The two UK undersea pump stations are complex underground structures situated at low points some 6 and 16 km from the Shakespeare Cliff adit. Fig. 22 shows the layout of these structures built mainly in machined, cast iron, bolted and grouted linings, with the NATM used for the principal shafts and intersection chambers.

79. The pump stations are significant tunnelling works in their own right, but building them entirely from a single-track possession in the service tunnel while servicing the TBM

drive and the crossover tunnel beyond was a major logistical feat.

80. The tunnels were constructed mainly by traditional hand tunnelling techniques, cutting the ground with pneumatic tools and erecting linings by winch. The Chalk Marl and the Glauconitic Marl below have compressive strengths more than 6 times that of London Clay. The 0·75 m long ductile iron rings have machined joints and grooves for neoprene seals.

Crossover

81. Two undersea crossovers are provided at the third points in the tunnels. They permit trains to cross between the running tunnels, and they allow closure of sections of the track for maintenance purposes. The UK crossover shown in Fig. 23 is some 8 km from Shakespeare Cliff in an area of minimum faulting.

82. The construction programme required the crossover to be completed before the arrival of the running tunnel TBMs. A multidisciplinary development study carried out

Fig. 21. Opening set

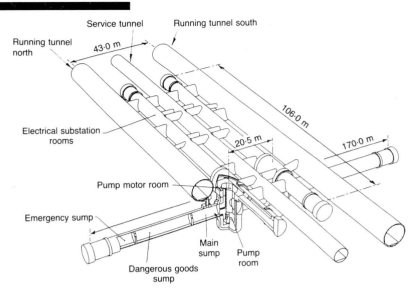

Fig. 22. UK pumping station

early in 1987 considered many alternative arrangements and concluded that this would best be achieved by the use of a single crossover chamber incorporating a scissor track crossing. The two running tunnels were deflected from the standard 30 m centreline spacing to 10·5 m centres at the crossover head walls. The service tunnel was diverted approximately 40 m to the north of the cavern. The

design and construction of the cavern are based on the principles of the NATM. Owing to the large size of the cavern, great care was taken during the design reviews and risk assessments of the excavation and primary support stages of the scheme selected.

83.   All construction work for the crossover was carried out from a two-branch adit extending from an enlargement of the marine service tunnel.

84.   The lower branch gave access to two sidewall headings, and the upper branch gave access to the top arch. The chamber was excavated in five main stages: two side wall headings; a top arch; the centre section; and the invert. The sequence was to drive the two side wall headings first, followed by the top arch. This sequence enabled exploratory probing from the side wall headings before excavating the top arch which if necessary could have been carried out in stages. Excavation and primary lining for the complete 164 m × 21 m × 15·4 m chamber and all access adits from the service tunnel took 15 months. Roof settlement was carefully controlled and was generally of the order of 45 mm. On one occasion, owing to unpredicted water pressure, settlement reached 70 mm. This situation was quickly stabilized by drilling pressure relief holes.

Fig. 23. UK undersea crossover

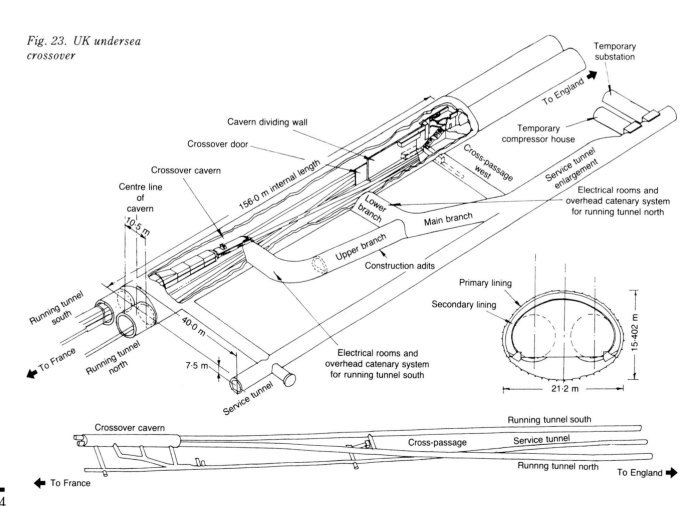

85. The contract programme was based on the completion of the excavation and primary lining by the time the running tunnel TBMs arrived. In the event the TBMs were delayed by $3\frac{1}{2}$ weeks. This delay was caused by the need to keep the service tunnel driving, which resulted in an overstretching of the back-up services. In addition it was difficult to control temperature and humidity with a large number of working faces all ventilated via the marine service tunnel. The resulting conditions were difficult for both men and machines.

86. There is no doubt that the construction of this large chamber so far offshore using the NATM is a step forward in the use of this tunnelling technique. The running tunnel TBMs broke through the end wall of the chamber, and were pushed through the chamber in approximately one week, using the bottom hydraulic rams pushing against sacrificial tunnel segments. The machines were then relaunched at the other end of the chamber. Fig. 24 shows the south TBM proceeding through the crossover. The secondary lining, varying between 600 and 700 mm thickness, was placed using a 5·5 m long steel-lined shutter. The final civil works involved the construction of central dividing walls at each end of the chamber. The 70 tonne doors for operational use will slide back

against these walls.

87. The French crossover was constructed using an entirely different technique. This consisted of driving a number of parallel adjacent headings approximately 3 m wide and filling these with concrete. When completed, this formed the permanent crossover walls. It can be seen from Fig. 25 that this technique minimized the exposed ground at any one time. The method had been used before on the Mount Baker road tunnel in Seattle.

Fig. 24. Running tunnel TBM in crossover

Fig. 25. French crossover

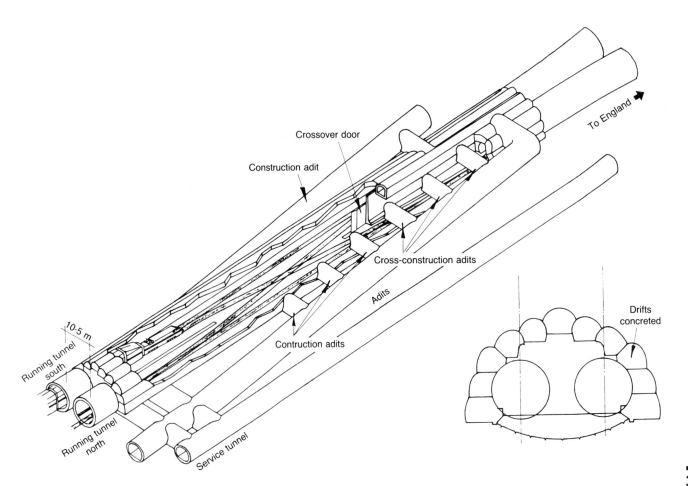

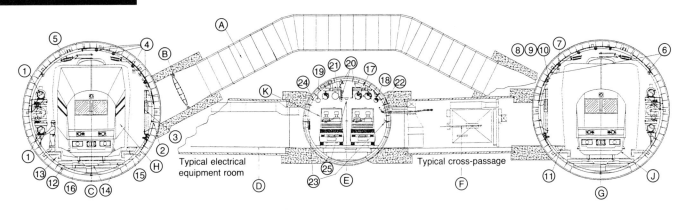

Running tunnel north marine   Marine service tunnel   Running tunnel south marine

A Piston relief duct at 250 m intervals
B Piston relief damper
C Running tunnel north marine
D Typical cross-passage with bulkhead door
E Marine service tunnel
F Typical technical room
G Running tunnel south marine
H Euroshuttle
J British Rail train
K Service tunnel transport system

*Fig. 26. Fixed
equipment layout*

Running tunnels
1 Cooling water pipes flow and return
2 Firemain 100NB (125 m each side of cross-passage)
3 Catenary tensioning weight (2 No. every 1·2km)
4 Tensioning weight pulleys every 1·2km
5 Catenary equipment (every 27 m)
6 Leaky feeder
7 Main lighting
8 2 × 20 kV cables
9 1 × 3·3kV cable
10 Low-voltage cables
11 Signalling cables
12 Phase 1 track concrete and drainage
13 Walkway concrete
14 Phase 2 trackbase concrete
15 Precast walkway units
16 Track blocks

Service tunnel
17 Drainage pipes 400NB 3 off + 1 (future)
18 Firemain
19 Firemain 250NB (future)
20 20kV/3·3kV and other supply cables
21 Leaky feeders
22 Control and communication cables
23 Main lighting push buttons (both sides)
24 Loudspeakers (both sides)
25 Main lighting

Fixed equipment installation

88. *General* The public perception of the project is the challenge of driving the tunnels. In reality the installation of the fixed equipment, in order to turn the tunnel into a railway system, has been equally challenging. The overall scope of the task can be appreciated from Fig. 26, which indicates the general layout of the fixed equipment.

89. *Installation* The installation work was carried out during and after the completion of tunnel driving. The objective is to have all installation and testing completed and ready for project commissioning during the last 6 months of the project period. Before project commissioning can take place, all testing, precommissioning and system commissioning has to be completed. The commissioning process is built around

(a) 50 primary systems (ventilation, drainage, power, rolling stock, etc.)
(b) 500 commissioning lots (a ventilation plant, a pumping station, etc.)
(c) 5000 test procedures (electrical, mechanical tests, performance tests, etc.).

90. The fifty primary systems are grouped into seven main system types.

(a) Control and communication: control centre, data transmission, radio, fire detection, signalling, terminal traffic management,

access control, C & C power supplies.
(b) Mechanical: cooling plants, tunnel cooling, tunnel drainage, normal and supplementary ventilation, tunnel fire fighting.
(c) Tracks: track in French terminal, track in UK terminal, track in tunnel.
(d) Power supply: grid connections, main HV sub-stations and traction supplies, terminals HV, LV systems and sub-stations, terminals exterior lighting, Sangatte shaft HV, LV and lighting systems, Shakespeare Cliff HV, LV and lighting systems, power system earthing, tunnel HV supply, tunnel LV and lighting system.
(e) Rolling stock.
(f) Networks: water treatment, potable water distribution, foul water drainage, sewage treatment, fire protection in terminals, fire detection network.
(g) Traction: overhead catenary system, tunnel earthing and bonding, rolling stock workshops, maintenance siding equipment, service tunnel transport.

91. The logistics of the installation work were very complex and required attention to the project programme. For simplicity of explanation the main phase of installation can be divided into pre-track work, the installation of the permanent track, and post-track work. All pre-track work was carried out from twin construction gauge track (900 mm). All post-track

work was completed from the standard gauge track (1435 mm) using possessions. Pre-track work included brackets, pipes and cables, all carried out as linear possessions. Post-track work included signalling, catenary, earthing and bonding and walkway construction. Cross-passages were fitted out both pre- and post-track.

92. All through the fixed equipment installation work the tunnel environment was sustained by the ventilation, drainage, water, power, communication and monitoring systems. Towards the end of the installation period these systems needed to be progressively decommissioned as the permanent systems were commissioned.

93. *Fixed equipment design* Fixed equipment design has followed a pattern of development studies, outline, definitive and detailed design with each stage being checked for safety and reliability by ET and the MdO. The fixed equipment works generally involve conventional designs with sub-contracts covering power supply, overhead catenary, tracks, control and communications and mechanical engineering sub-systems. The challenge has been the sheer scale of the work in a fast-track project environment. Initially the technical task was to provide sufficient fixed equipment information to allow the civil works to proceed. This was followed by extensive co-ordination to integrate the design with the civil works.

94. The power supply system takes supplies equivalent in capacity to a city the size of Leicester from Electricité de France at 225 kV and SEEBOARD at 132 kV. This is transformed to 21/3·3/0·4 kV for tunnel and terminal auxiliaries. The traction system utilizes a 25 kV single-phase overhead catenary which has to be balanced onto the incoming three-phase supply. This necessitated using the largest static compensation equipment ever installed in the world until now. Mechanical cooling, ventilation, fire fighting, pumping and drainage design have been co-ordinated with the civil works to ensure the correct equipment layout in the confined tunnel space allocations. Fibre optic transmission systems have been employed for the remote control of fixed equipment, and rolling stock control uses the latest 'in-cab' signalling, as used by the TGV in France.

95. *Tunnel track system* The major civil aspect of the tunnel fixed equipment, and of particular interest to civil engineers, is the tunnel track system shown in Figs 27 and 28.

96. The main design principles adopted for the tunnel non-ballasted trackform were

(a) the ability of the concrete support system to last 50 years under a potential traffic of 240 MGT/year (which will make it the world's most heavily trafficked railway)

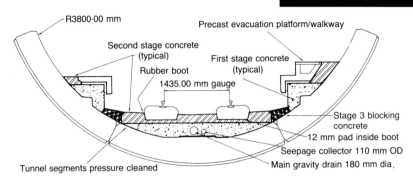

Fig. 27. Trackwork cross-section

(b) low aerodynamic resistance to the airflow between the track and the underside of the trains
(c) strict tolerances on gauge widening under combined vertical and lateral loads
(d) static and dynamic resilience of the system compatible with intense traffic loading
(e) good electrical resistance between the two rails
(f) a rail fastening which permits minor in-service adjustment of rail position
(g) economically maintained and renewed components.

97. The non-ballasted track system comprises pairs of independent concrete support blocks, spaced at 60 cm centres along the track, each block resting on a 12 mm thick micro-cellular pad inside a rubber boot and cast into the trackbase concrete. The rail, which is seated on an H-shaped rail pad, is fixed to the block by a fastening which comprises a hex-headed screw, phenolic washer, two spring steel blades, a nylon insulating clip and rubber O-ring. This assembly is 'bolted' into an insert cast into the concrete support block.

98. To ensure that the components of the system were fully compliant with the general criteria, a series of acceptance tests was performed. The major tests included those for static resilience, dynamic resilience with the application of 3 million load cycles, fastening

Fig. 28. Rail detail

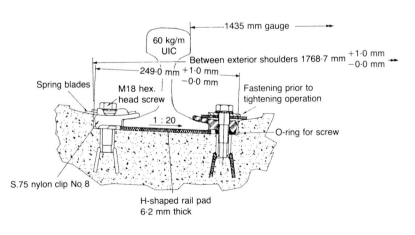

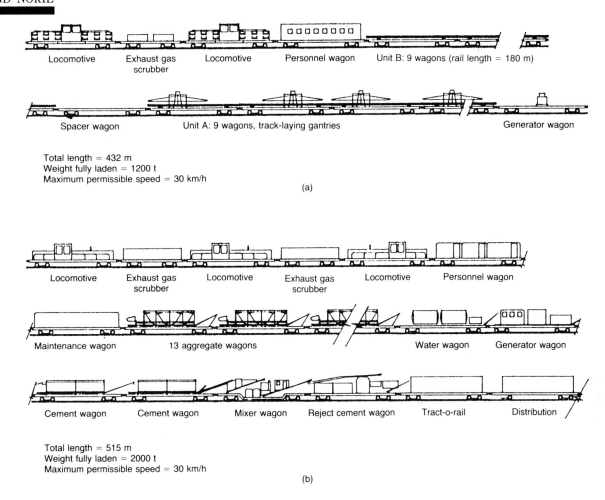

Locomotive Exhaust gas scrubber Locomotive Personnel wagon Unit B: 9 wagons (rail length = 180 m)

Spacer wagon Unit A: 9 wagons, track-laying gantries Generator wagon

Total length = 432 m
Weight fully laden = 1200 t
Maximum permissible speed = 30 km/h

(a)

Locomotive Exhaust gas scrubber Locomotive Exhaust gas scrubber Locomotive Personnel wagon

Maintenance wagon 13 aggregate wagons Water wagon Generator wagon

Cement wagon Cement wagon Mixer wagon Reject cement wagon Tract-o-rail Distribution

Total length = 515 m
Weight fully laden = 2000 t
Maximum permissible speed = 30 km/h

(b)

*Fig. 29. Track
installation:
(a) track-laying train;
(b) concrete train*

response tests, longitudinal restraint tests, and system fatigue tests with the application of 10 million load cycles. All the components of the system satisfactorily met all the criteria.

99. Since the track in the tunnel was considered to be potentially in a saline environment, there were also stringent anti-corrosion criteria to be satisfied. Extensive salt spray tests were therefore performed on the steel components of the fastening, utilizing various corrosion treatments. The bolts and inserts are treated with electrolytic zinc and passivated by bichromation and the spring blades are treated with fusion-bonded epoxy. This system in the assembled state proved to be satisfactory after a 1000 hour salt spray test.

100. The sequence of installation is shown diagrammatically in Fig. 29, and as follows.

101. The rails (which are supplied in 182·5 m lengths) are attached to the blocks, complete with all resilient and fastening components in the terminal area outside of the tunnel and stored as assembled rail strings.

102. The eight rail strings are taken into the tunnel on the track laying train and offloaded onto the stage 1 trackbase by a series of rail gantries. This activity can progress at a rate of

730 m of track per day from the tunnel portal. Temporary gauge bars are attached permitting the construction trains to run over the track in its temporary position.

103. After a predetermined length of track (of the order of several kilometres) has been laid out, the formwork for the first stage of the walkways is set up, and reinforcement is placed and concreted.

104. The track is then aligned and levelled to its correct three-dimensional position by means of a Framafer track lifting and lining machine and confirmed by a CM10 track recording device. The track is supported in its correct position by screw jacks, both vertically and horizontally.

105. The stage 2 trackbase concreting operation then commences. Aggregates are dry-batched on the terminal site and loaded onto the concrete train; mixing with cement and water is an onboard activity. Concrete is loaded from the end of the train onto hopper gantries which run on pneumatic tyres on the first stage walkways. The concreting of the stage 2 trackbase proceeds at approximately 300 m per day from the tunnel portal.

106. The precast walkway units are taken

into the tunnel and positioned by means of a special handling train.

107. Finally, the infill concrete is placed behind the walkway units, and the stage 3 trackbase or 'blocking concrete' is placed between the edge of the stages 1 and 2 trackbase and the in situ walkway concrete.

Construction safety

Safety organization and methods

108. TML's Health and Safety Department provided advice on occupational health and safety matters, operated the on-site medical facilities, and controlled the fire and rescue team. The management of safety was carried out by the Construction Department.

109. Working as a team, the members of the Health and Safety Department, were able to deploy a wide range of skills and experience in support of the line management's discharge of it's duties under current health and safety legislation.

110. In accordance with the recommendations made following the public inquiries into recent major accidents, e.g. the capsize of the *Herald of Free Enterprise*, the London Underground fire at King's Cross Station, and the fire and explosion on the offshore installation Piper Alpha, the philosophical concept of proactive safety management has been embraced, and continues to be developed within the Project Management team.

111. The effect of these events and accidents on the project and on the development of the safety culture within the project has been to focus line managers' attention on the safety element of their responsibilities. A safety plan, the TML Safety Management System comprising six complementary and interactive elements, was defined and implemented. This provided a framework within which each manager could positively address the safety matters affecting both the work and the operatives for whom he was responsible, measure the effectiveness of actions taken, identify potential, and propose remedies.

112. The key to safety lies in the knowledge of the correct method of work, and a significant amount of time is devoted to training, the aim being to produce a better-qualified workforce within TML, able to use judgement and forethought to enable timely and appropriate action to be taken to ensure a safe work environment.

Practical safety activities

113. A number of practical steps were taken to achieve the major attitudinal changes upon which an improved safety performance was based. Tiered monthly meetings dedicated solely to safety were held, the most senior chaired by the Chief Executive. At all tiers line managers had to report on the safety performance of their group of employees and the remedial actions they proposed to avoid recurrence. The tiered pattern was used to pass information and ideas both up and down the organization.

114. All accidents were investigated by line managers. At the minor accident end of the spectrum the investigator was the injured party's chargehand. The two principal elements addressed in all the investigations were the cause and prevention of recurrence. The philosophy adopted was that all accidents could be avoided.

115. A series of mostly one-to-one audits were carried out by foremen, supervisors and chargehands on operatives at work. Their style—achieved by on-the-job training—was unique and well outside the norm for the construction industry; they were 'no-risk' discussions. The auditee was encouraged to tell the auditor of any of his observed actions which were less safe than they might have been. The auditor structured the discussion to the acronym PHOTO (personal protective equipment; housekeeping; orientation; tools; and operating procedures). The auditees were also encouraged to make safety suggestions. Audits took about 15 minutes and were written up by the auditor after the audit and out of sight of the auditee. Completed audits which identified the auditor but not the auditee were analysed by a senior line manager and summarized for the local line manager to indicate trends and actions required. The majority of auditees responded very well to the no-risk audits and by volunteering self-improvements, were far more likely to change bad habits than they would have done in response to the more normal, somewhat robust, construction industry supervisory style. Many, previously unrecognized, problems were identified and dealt with by the collection and collation of sound data emanating from the workforce. The attitudinal change on the auditors was also pronounced. About one-eighth of the workforce was so audited each month.

116. For the sake of the associated publicity, safety prizes were awarded by lottery to an individual member of groups of various sizes who had achieved 25 000 hours of totally accident-free work activity.

117. A number of poster campaigns were staged; the best of these were cartoons. The artwork was done by two amateur, talented, hourly paid operatives to whose cartoons everybody was able to relate as they were so clearly in-house and apposite. Such campaigns addressed specific issues such as track safety, eye protection and hearing, and they were complemented by other media such as written notes attached to payslips, on-site video loops and special talks to all employees all timed to coincide for maximum impact. One of the artists—

whose normal duties involved driving Alimak hoists in the Shakespeare Cliff access shaft — contributed the artwork to a humorous booklet entitled 'Charlie the chargehand' whose serious aim was to address the required roles for sound chargehand performance.

118.   Chargehand and supervisory roles, and principally their safety roles, were covered twice in the project at a round of special seminars chaired by senior line managers able to relate to the chargehand ethos. Some 600 people attended each of these seminars, twenty or so at a time, which were very informal, lasted two hours or so, and gave rise to many varied and valuable safety suggestions, almost all of which were put into practice within days of that seminar.

119.   The realization by line management that the chargehands and their workforce were an abundant and valuable source of safety talent was another example of a crucial attitudinal change. In some cases this talent was tapped formally, and hourly paid operatives sat on a very small *ad hoc* committee, which (among other tasks) tackled revisions to procedures. Such revised procedures always led to productivity and safety improvements, and (of greatest importance) were accepted by those doing the work as being practical and easy to follow.

120.   An attempt to obtain safety improvements suggestions by completing a potentially anonymous pro forma was not a success; TML operatives preferred to make their suggestions face-to-face in informal discussions with their line management team.

121.   Each month each operative received about four field safety talks (FST); the topics chosen —usually of a safety nature— were catholic. Although the choice of topic was usually that of the manager or supervisor, many were based on the informal discussions which followed the previous FST.

Control room

122.   *General*   The control room activities were a vital part of the tunnel construction operations. The control room acts as both a communications centre and an early warning indicator for various construction activities in the tunnel.

123.   *Construction systems monitoring* Sensors situated at numerous positions continually sample the environment. Oxygen, carbon monoxide and temperature are checked throughout the tunnel complex; flammable gas, nitrogen dioxide and carbon dioxide are checked less frequently. Information on all these is fed back to the control room where it is converted into a series of visual displays which can be shown on any one of six screens. A team of four operators continually monitors these screens for potential environmental problems.

The electrical, pumping and ventilation systems are also monitored from the control room, from where the fans are also controlled.

124.   All incidents, accidents and emergencies are reported in the first instance to the control room. From here the appropriate course of action is taken. This may result in sending a person for urgent treatment. All of these are co-ordinated throughout the control room where a senior manager (main controller) is available 24 hours a day, 7 days a week, to make any crucial decisions.

125.   Communications between underground areas and the surface are routed through the control room in a variety of ways, including a public address system which can reach all parts of the tunnel system.

126.   *Emergency procedures*   All emergencies are co-ordinated through the control room which acts as a vital link between surface and underground activity. The site has its own rescue centre and medical centre, from where staff can be despatched at short notice to tackle any emergency from fire to entrapment or injury. In addition, these resources provide valuable backup to the emergency services for more serious emergencies. The main controller co-ordinates the initial response until the emergency services' co-ordinators are in place in the communications centre. A set of procedures has been evolved to deal with a typical emergency; this has 23 guideline appendices for the control of specific types of emergency. A further procedure covers any bi-national emergency, in which teams from both sides of the Channel are required to work together.

Progress

127.   Design work started on the project in June 1986, and the TBM drives started in December 1987. Within that 18 month period the marine service tunnel TBM had been designed, was manufactured, and was driving the tunnel. A precast factory which was to be the biggest in Europe, and probably the world, had been established on a virgin site at the Isle of Grain and was in production. Site investigations had been carried out in the Channel, plant and materials procured, tunnelling work was well underway at Shakespeare Cliff, and sufficient design work had been done to allow all this to happen.

128.   The TBM drives were completed in June 1991. This meant that over 150 km of tunnel (most of it undersea) between two countries, including much of the design, equipment manufacture and procurement, were completed in 42 months.

129.   The marine service tunnel TBM, the first on the project to be commissioned, cut rock on 29 November 1987. The other five UK TBM drives started progressively between August 1988 and November 1989. At peak, five

TBMs were in operation concurrently between August 1989 and September 1990. Each drive was manned on a three eight-hour shift basis, seven days per week. Shift changes took place at the TBM.

130. Progress on each of the marine drives was slow in the initial stages, as the TBMs encountered wet and badly broken ground. Rates of advance accelerated as better ground conditions were met and working methods and equipment were refined, and the early delays to the programme were gradually recovered. Over the final six months of the marine running tunnel drives, the work forged ahead of programme as rates of progress of over 350 m per week were consistently achieved. New production records were repeatedly established through the period, culminating in an advance of 428 m by the running tunnel south marine TBM in the week ending 24 March 1991.

131. The breakthrough of the running tunnel south marine tunnel in June 1991 marked the completion of the tunnel drives. By that date a total of 84 km of bored tunnel had been driven by the UK side, and 3.9×10^6 m³ of rock had been excavated.

Quality assurance and quality control

132. The quality system established for the Channel Tunnel project as a whole includes a classification system which allows flexibility, with the application of the requirements depending on the criticality of the work, and stresses the need for a planned approach to the achievement of quality.

Table 1. Main tunnel statistics (UK side)

Civil construction					Fixed equipment	
Permanent works		Main construction plant			Permanent works	
Description	Quantity	Description	Quantity	Value	Description	Quantity
TBM driven tunnels	84 km	*Tunnels*			*Mechanical*	
		TBMs (cw spares)	6	£60 m	Pipework, ave. 335 mm	240 km
		Roadheaders	5	£2 m	Valves	2500
Total tunnel spoil	3.90×10^6 m³	Electric locos			PRD dampers	130
		Rack	18	£6.1 m	Cross-passage doors	270
TBM tunnels spoil	3.61×10^6 m³	Adhesion	60	£12.1 m		
		Diesel locos	58	£7.0 m	*Catenary*	
Crossover spoil	48 000 m³	Diesel manriders	16	£1.9 m	Wire	250 km
		Muck cars	350	£9.1 m	Insulators	6000
Cross-passages		Flat wagons	660	£6.6 m		
Number	245				*Fitted out rooms*	
Length	2.76 km	*Conveyor system*	System	£10.1 m	Electrical	98
		Rail track: km	200	£5.5 m	Sub-stations	18
Piston relief ducts		550 V pickup: km	55	£4.4 m	Evacuation passage	178
Number	106	Ventilation duct: km	80	£2.6 m	Signalling	32
Length	2.28 km	Ventilation fans	109	£7.9 m	Pump stations	2
		Power cable: km	250	£4.5 m		
Grout volume	60 000 m³	*Water and drainage*			*Electrical*	
		Pipe: km		£4.7 m	Cabling	600 km
PCC rings		Portal cranes	7	£2.2 m	Supports	150 000
Number	53 150	Crawler cranes	10	£2.4 m	Cable tray	250 km
PCC segments		Desalination plants	2	£3.5 m	Lights	13 000
Number	442 000					
Weight	1.6×10^6 t	*Precasting works*			*Control and communication*	
		Carousels (Sacma)	8	£3 m	Cabling	275 km
SGI rings (750 wide)		12.5 t portal cranes	17	£1.8 m		
Number	17 000	12.5 t EOT cranes	15	£0.4 m	*Track*	
Weight	49 400 t	Concrete moulds	599	£7 m	Rails	120 km
		Concrete batchers	4	£1.7 m	Sleeper blocks	100 000 prs
Shotcrete volume	16 000 m³	Tolerance testing machine			Fastening kits	400 000
Concrete volume	150 000 m³	(Crown Winley)	2	£0.2 m		
		Tractor units	10	£0.3 m		
Sea wall	1700 m		4			
		Flat trailers	20	£0.13 m		
Pump station excavation	7757 m³					

Data refer to 84 km of UK tunnels. In addition there are 63 km of tunnel on the French side.

133. Two separate quality roles are in operation for the tunnelling works, the quality assurance (QA) function and the quality control (QC) organization. The QA function establishes the system that provides a quality product and monitors its implementation. The QA department has a direct reporting line off-site to higher management, dealing with quality planning, surveillance of the product, review and audit. The QC organization is part of the UK Construction Group and provided full-time inspection cover of tunnelling works to ensure compliance with the specifications and drawings.

134. The QC role is similar to a resident engineer's role on a traditional construct-only contract which has an independent engineer. The QC department reported to the TML engineering department, and was also responsible for inspecting and certifying that the civil tunnel works were of adequate standard before handover for the installation of the mechanical and electrical services.

Conclusion

135. It seems likely that this great project will open not long after the originally projected time. The work on-site will have been carried out in a six year period. The scale and success of the achievement, in terms of the design task, the logistics and the technical challenge below ground, set against the time frame, mean that all personnel involved in the project should be proud to have been involved. The shear scale of the project is illustrated in Table 1. Although the gestation period of the project has actually been 200 years, the execution has been rapid. The timing has worked out propitiously in terms of both the state of the art of tunnel boring machines, the modern survey techniques involved, and perhaps most important of all the climate of industrial relations in the UK. There appears little doubt that the great effort put in by client, contractor, funding agencies and all others involved will have been worthwhile, and that the tunnel will play its part in the economic health of the United Kingdom.

Acknowledgements

136. The Authors are grateful to their TML colleagues who contributed and gave advice, and to Eurotunnel and TML, for permission to publish this Paper. In particular, they acknowledge the contributions made by Keith Wilkinson, Safety Director, for the section on construction safety, and by Colin Bayliss, Engineering Manager, for the section on fixed equipment installation.

References

1. CRIGHTON G. S. *et al*. Supplementary site investigations for the Channel Tunnel 1986–87. *Proc. 5th Int. Symp. Tunnelling '88*. Institution of Mining and Metallurgy, London.
2. HESTER J. C. *et al*. Channel Tunnel UK–France: the UK TBM drives. *Rapid excavation and tunnelling conference*. Institution of Mining and Metallurgy, London, 1991.
3. CRIGHTON G. S. and LEBLOND L. Tunnel design. In *The Channel Tunnel*, pp. 95–135. Thomas Telford, London, 1989.
4. BIGGART A. R. *et al*. UK tunnels construction. In *The Channel Tunnel*, pp. 231–248. Thomas Telford, London, 1989.
5. BIGGART A. R. *et al*. The Channel Tunnel project: the UK tunnels—a design and construction overview. *Tunnelling '91 Conf*. Institution of Mining and Metallurgy, London, 1991.

Geology, alignment and survey

P. Varley, PhD, MSc, BSc, MIMM, A. Darby, MA, MSc, MICE, FGS, and
E. Radcliffe, MBE, FInstGES

Proc. Instn Civ.
Engrs, Civ. Engng,
Channel Tunnel,
Part 1: Tunnels
1992, 43–54

Paper 9932

■ The Paper describes the regional and structural geology of the Dover Strait area, the modern site investigation techniques used to identify the preferred tunnelling medium and to determine the boundaries and condition of the chosen stratum, and the collection and use of geological data during construction. The establishment of the tunnel alignment for a high speed railway system is described in terms of the reconciliation of operational and construction criteria and spatial constraints within the basic geological structure. A reference alignment was defined, which was then partially revised in the light of additional geological data, design development and actual driving experience as the work progressed. Finally, the Paper covers the establishment of horizontal and vertical control survey networks, based on an appropriate projection and grid, and a common elevation datum, the transfer of survey control underground, and the survey techniques and instruments used to deal with the problems encountered in maintaining accurate drive alignments.

Part 1: geology

Regional and structural geology of the Strait of Dover

The Strait of Dover cuts through the northern limb of a major, east–west aligned anticline, of Tertiary (Alpine) age (Fig. 1). This 'Weald-Boulonnais' anticline consists of Jurassic and Cretaceous strata (notably the Chalk and underlying Gault Clay), folded gently over a structurally complex basement of Palaeozoic rocks. On the UK side, the Cretaceous strata have a gentle north to north-east dip of usually less than 5°, although towards France this increases, locally, up to 20°. Superimposed on this basic structure are a series of minor east–west anticlines, one of which has produced a small inlier of Gault Clay to the south of Dover. Similarly, at Quenocs on the French coast, the southern limb of the anticline has developed a relatively sharp dip towards the south-east.

2. Deposition and subsequent erosion of the Palaeozoic (Carboniferous) and Mesozoic strata were dominated by the presence of a land mass called the London–Brabant Massif. Structures within this massif also controlled the response of these strata to deformation by the main tectonic events, including the subsequent formation of the Weald–Boulonnais anticline. In consequence, the faults present in the Cretaceous strata reflect those in the Palaeozoic (visible in the inliers of the Boulonnais and in the Kent Coalfield).

3. The geological structure along the route is relatively simple and there are no major displacements. The largest faults occur on the French side: at km 47·0 (8–10 m downthrow south) and km 49·6 (3–5 m downthrow north), aligned (with other more minor faults) with the axis of the main anticline. The easterly dip component has caused the Chalk/Gault interface to fall to a low point about 34 km from the English end. From here to the French side there are several further low points, resulting from the degree of secondary folding, which, as with the faulting, is greater on the French side and decreases towards the UK (Fig. 2). The basic stratigraphy across the Strait is given in Table 1.

4. The Chalk strata increase in clay content and decrease in calcium carbonate content with increasing stratigraphic depth, changing colour from the characteristic white chalks seen in the cliffs to a grey chalky marl. A thin transitional stratum, rich in the mineral glauconite, separates these from the underlying Gault Clay. This gradational change in colour is accompanied by a change in rock properties. The carbonate-rich rocks are stronger but more brittle and do not deform without fracturing as readily as do the more 'plastic', clay-rich rocks. They would, in consequence, be expected to transmit more water than the underlying strata. Conversely, the Gault Clay has a greater potential for immediate and long-term swelling, and requires a stronger lining than the strata above it.

5. The lower part of the Lower Chalk is recognized as a distinct sub-unit, called the Chalk Marl, which consists of cyclic alternations of grey marly chalk with stronger limestone bands. This sub-unit possesses the best characteristics of both rock types.

Site investigation for the Channel Tunnel

6. The logistical demands controlling construction and use of a marine tunnel increase naturally as its length increases. From the earliest proposed scheme, routes for a Channel Tunnel have therefore been selected from a narrow corridor of sea, where the separation between Britain and France is at a minimum.

P. Varley,
Geotechnical
Design Manager,
Translink Joint
Venture

A. Darby,
Design Team
Leader—Alignment,
Mott MacDonald
Group

E. Radcliffe,
Chief Surveyor—
Tunnels,
Translink Joint
Venture

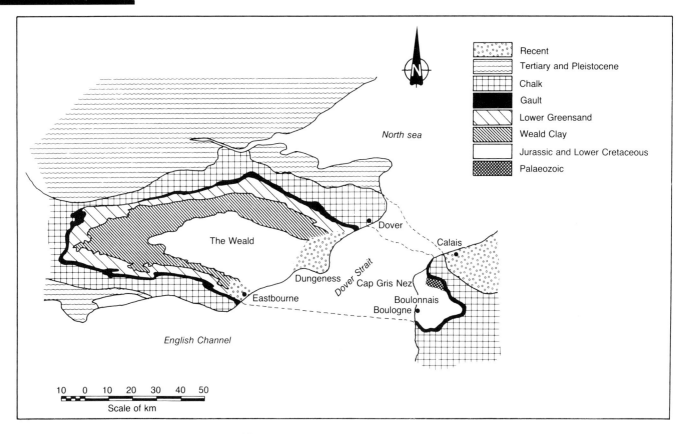

Fig. 1. Geological structure of the Strait of Dover

Fig. 2. Longitudinal section of service tunnel showing geological controls on the alignment

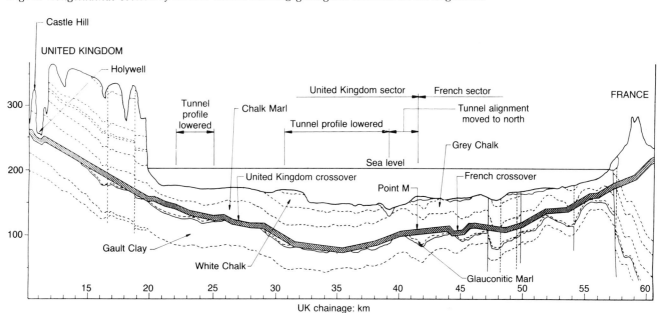

This corridor, across the Strait of Dover, is bounded to the south by a line from Folkestone to Cap Gris-Nez and to the north from St Margaret's Bay to Sangatte.

7.   As early as 1628, a writer called Verstegan noted the similarity between the rocks forming the cliffs on either shore and postulated that at some point in the past they had been joined by dry land. This was confirmed about 250 years later by stratigraphic and palaeographic comparisons. Samples were collected from the seabed and, after much debate, the importance of the Chalk Marl was recognized as a good tunnelling medium. This was eventually confirmed in 1880–83 with the drivage of over 3 km of tunnels from either shore, using Beaumont-English tunnelling machines. While geological maps and sections were produced, it had still to be assumed that the strata visible in the opposing cliffs extended in an unbroken arc under the Strait.

8.   The first modern investigation for the route for a fixed Channel crossing was undertaken in 1959. About 1000 line kilometres of geophysical (seismic reflection), traverse lines were run, in conjunction with marine and land boreholes. The geophysical survey gave some

Table 1. Simplified geological succession

Formation	Typical thickness: m		Description
	UK	France	
Middle Chalk	80–90	60–70	White Chalk, flints at top Plenus Marl at base
Lower Chalk	75	70	Top: White Chalk Middle: Grey Chalk—marly chalk Lower: Chalk Marl—chalky marl Base: Glauconitic Marl—marly glauconitic sandstone (2–3 m)
Gault	45	15	Calcareous, overconsolidated clay
Lower Greensand	50	15	Weakly cemented sand and clays

information on the shallow structure and faulting over the Folkestone/Cap Gris-Nez to Dover/Sangatte corridor (Fig. 3), and suggested that deep, drift-filled channels would not present an unavoidable hazard to the tunnel. It was used to select a route which was then examined in much greater detail by a major geophysical and

Fig. 3. Marine geophysical surveys

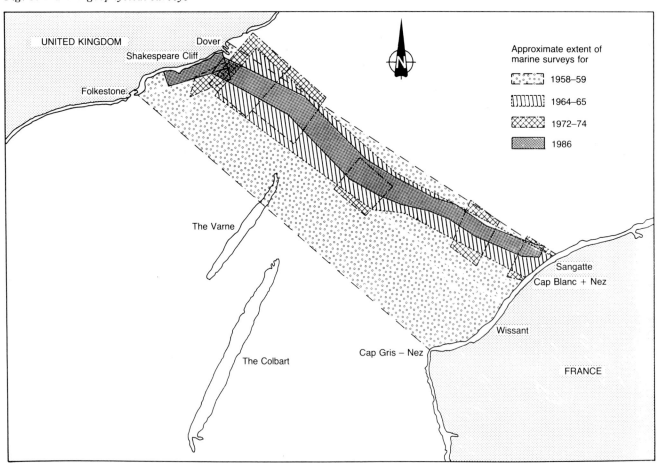

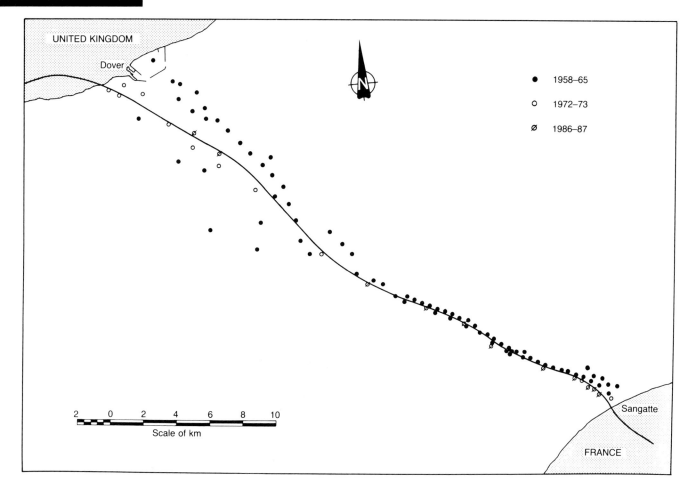

UNITED KINGDOM

Dover

● 1958–65

○ 1972–73

ø 1986–87

2   0   2   4   6   8   10

Scale of km

Sangatte

FRANCE

*Fig. 4. Borehole
locations in the
Channel*

borehole site investigation in 1964–65 (Fig. 4).

9.　The 1964–65 campaign located a zone of
deep weathering associated with the former
seaward extension of the Dour River Valley,
through Dover Harbour. A further campaign of
geophysical surveys and boreholes in 1972–73
sought to increase the level of knowledge on
this, and other specific areas of interest highlighted by the preceding investigations, and to
find more competent tunnelling ground. The
tunnel was realigned on the UK side of the
Channel, moving the tunnel towards the south.
A major infilled valley system, called the Fosse
Dangeard, which reaches a depth of up to 80 m
below seabed level, was discovered in the
Strait. This valley is believed to have developed by river erosion of the Gault Clay sub-
crop during the Quaternary, and is to some
extent fault controlled. A relatively shallow
(10 m deep) arm of the Fosse crosses the alignment.

10.　The contract awarded to TML included
the requirement for a further site investigation,
consisting of a geophysical survey along the
corridor of the chosen alignment, marine and
land boreholes. The borehole data were used to
calibrate the geophysical surveys to identify
the following horizons: base of alluvium; base
of Glauconitic Marl; base of Gault Clay (top of
Greensand).

11.　Seabed levels were available from a
separate bathymetric survey. In order to estimate the levels of the horizons on various
tunnel alignments, the geophysical results were
entered in a geostatistical program which predicted spot levels on a 20 m by 40 m grid along
a 1 km wide corridor between England and
France. A final geophysical survey was undertaken in 1988, specifically for the crossing
beneath the Fosse Dangeard and the sites of the
UK and French crossovers.

12.　Table 2 shows the general geotechnical
characteristics of the strata, as defined by the
boreholes drilled for the UK crossover. The
detailed subdivision of the Chalk Marl into
three zones with different physical properties
was unknown before the 1988 investigation.
The recognition of zone 6A, at the top of the
Gault Clay, was important for the vertical location of the crossover. The significantly better
mechanical properties of this zone, imparted by
a higher carbonate content, permitted excavation of the base of the crossover through the
Glauconitic Marl and into the upper zone 6A,
thereby increasing the cover between the roof
of the crossover cavern and the seabed.

13.　The large number of boreholes sunk,
particularly on the French side of the Channel,
became a constraint to the alignment owing to
their uncertain location and backfilling. The

Table 2. Strata characteristics at UK crossover—based on phase II (1988) investigations

Description	Average thickness	Typical intact UCS: MPa	Undrained Eur* stiffness: GPa	Typical mass permeability: m/s
Grey Chalk (unweathered)	5	7·5	1·65	$>2 \times 10^{-8}$
Upper Chalk Marl	20†	8·5	1·65	2×10^{-8} to 4×10^{-7}
Lower Chalk Marl	8	3·5	0·8	1×10^{-7}
Basal Chalk Marl	4	7·5	1·5	7×10^{-8}
Glauconitic Marl	2·5	13	5·0	5×10^{-8} to 5×10^{-10}
Zone 6A (upper)	4	8	2·1	1×10^{-8}
Zone 6A (lower)	4	3·5	0·8	1×10^{-8}
Gault Clay (upper)	17	2	0·45	1×10^{-8}

* Eur: modulus of elasticity determined from unload–reload cycle prior to failure.
† Top of Chalk Marl based on geophysical reflector; top of Bed G in Amedro's biostratigraphical classification about 5 m lower.

accuracy of each location survey was assessed and circles of maximum possible error drawn around each given location. The possibility of intersecting uncharted boreholes, drilled for other purposes than the Channel Tunnel, could not be discounted, and the tunnelling machines were therefore designed to cope with any such unexpected intersections.

14. Implications of the geology for the tunnel alignment are discussed in §§ 32–42.

15. The portal for the Channel Tunnel on the UK side is located within a major slip, on the western flank of Castle Hill. An extensive array of borehole instruments, to monitor the stability of the escarpment and main Castle Hill slip as the work progressed, was installed during the 1986–87 site investigation and in

stages during the subsequent construction of the tunnel.

Data collection and use during construction

16. Overlapping destructive probe holes were drilled from the face of the marine service tunnel, to give advance warning of unstable ground and high or saline water inflows. Micro-palaeontological examination of the returns was used to determine the relative position of the face within the Chalk Marl. This was also monitored with regular vertical cored holes drilled through the tunnel invert (Fig. 5).

17. Detailed geotechnical logs were made of the ground exposed in the side walls and (where possible) in the face, and these data

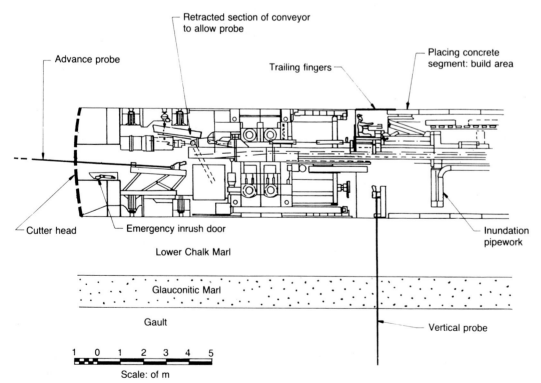

- Retracted section of conveyor to allow probe

- Advance probe

Trailing fingers

- Placing concrete segment: build area

- Cutter head

- Emergency inrush door

Lower Chalk Marl

Glauconitic Marl

Gault

- Inundation pipework

- Vertical probe

Scale: of m

Fig. 5. Arrangement of probing rigs in service tunnel

Table 3. Applied cant and cant deficiency*

	Design speed: km/h	Radius: m		
		4200: mm	5000: mm	6000: mm
Applied cant		50	42	35
Cant deficiency	200	63	53	44
Cant deficiency	160	22	19	15
Cant deficiency	130	−2	−2	−2
Cant deficiency	100	−22	−18	−15

* Rates of change of cant and cant deficiency were selected to suit the shuttle rolling stock

Desirable rate of change of cant	35 mm/s
Desirable rate of change of cant deficiency	35 mm/s
Maximum rate of change of cant	37 mm/s
Maximum rate of change of cant deficiency	37 mm/s

were used to predict ground conditions in advance of the excavation of the running tunnels and cross-passages.

18. Two faults with minimal associated disturbance were predicted to intersect the land drives, and these potential hazards were covered with probe drilling from the face of the land service tunnel.

Part 2: alignment

19. Design criteria were key parameters, affecting not only the alignment of the tunnel but also the design of the rolling stock and locomotives, and the requirements for construction plant and methods. The criteria were agreed with Eurotunnel in 1987 and were subject thereafter to only very minor changes. They are considered below.

Operational criteria

20. The Concession Agreement specified that the design should allow for the installation of railway track in both running tunnels suitable for trains travelling at speeds of 200 km/h. The alignment of the track was designed first, and the alignment of the tunnel was determined subsequently by adjusting for the effect of cant of the track. The track design was a compromise between the requirements for fast passenger-only trains, shuttle trains susceptible to sway on account of their large section, and comparatively slow freight trains which could cause excessive wear on the low rail on curves if cant were applied to suit the faster trains.

21. A preferred minimum radius of 6000 m on horizontal curves was requested by Eurotunnel to minimize the conflicting requirements, although the absolute minimum radius remained at 4200 m as specified in the Concession Agreement. The transitions between the circular curves and straights are Crisswell clothoids. The applied cant and cant deficiencies

are listed in Table 3. The vertical curves are parabolas with radii at the vertex of 15 000 m, resulting in a vertical acceleration of 0·03g at the design speed of 200 km/h.

22. The spacing of the running tunnels was fixed at 30 m, with the service tunnel midway between them. This allowed adequate space for the equipment rooms between the service and running tunnels. The interaction of adjacent tunnels imposed only minor effects on the lining design.

23. The terminal at Cheriton and the existing length of service tunnel under Shakespeare Cliff provided two fixed points between which the grade of 1·09%, was satisfactory for train operation. A steeper gradient would have slowed trains travelling uphill, and 1·1% was adopted as the maximum. The minimum gradient at any section was controlled by the drainage flows from predicted seepage and the maximum drain size that could be fitted into the trackbed. An absolute minimum grade of 0·18% was chosen to achieve a satisfactory drainage regime. At low points in the tunnel profile, it was necessary to provide pumps. It was desirable to minimize the number of pumping stations both on account of their cost and to reduce maintenance.

24. Only rubber tyred vehicles will be used in the service tunnel, with a maximum speed of 80 km/h, such that operational criteria were much less stringent. The desirable minimum horizontal and vertical radii were 1000 m and 3000 m respectively. The maximum gradient was 3·5%.

Geological criteria

25. The optimum tunnelling horizon is within the unweathered Chalk Marl, 5–15 m above the base of the Glauconitic Marl. The Chalk Marl had proved suitable for rapid excavation by machine during the 1975 works and was found generally to have good short-term stability in tunnel excavation.

26. The minimum cover to the base of the superficial deposits was set at 18 m. In the UK sector, the tunnels were kept at least 10 m below the base of weathering, as identified from the boreholes.

Spatial constraints—land section

27. The Folkestone terminal was located alongside both the M20 motorway and the railway line to London. At Shakespeare Cliff, on the Kent coast, the new works were to incorporate the access adit and 432 m length of service tunnel excavated in 1974–75. These constraints fixed the line and level of the works at Shakespeare Cliff.

28. Some scope remained for adjustment of the portal positions at Castle Hill to minimize the effect of the landslips here. Within Holywell, the route for the cut and cover works was

selected following detailed discussion in the House of Lords' Select Committee on the optimum balance between scientific, landscape and nature conservation issues. Once the Bill for the link had passed through Parliament, any variation of the alignment in the UK land sector was restricted to 100 m either side of the proposed tunnel alignment.

Construction criteria

29. The minimum horizontal radius of the service tunnel was set by the TBM gantries, which were designed to negotiate curves of 4000 m nominal radius. Rail haulage, used during construction, required a limiting gradient of 1·1% for the service tunnel, the controlling factors being locomotive power and adhesion.

30. Construction required local changes to the alignment criteria. The UK undersea crossover cavern was excavated from the service tunnel, and its alignment was lowered to pass under running tunnel north and was offset 45 m from the centre-line of the cavern, to provide space for the construction adits linking the tunnel and cavern.

31. The standard separation of the running tunnels was widened at the undersea pumping stations, so as to allow the pump motor rooms and shafts to be excavated between the running and service tunnels. At km 25·0, this increased the separation to 38 m, and at km 34·8, where the greater depth required thicker linings, the separation was increased to 43 m.

Reference alignment

UK underland

32. The UK underland alignment was controlled by the spatial and operational criteria described previously. With the alignment fixed at Castle Hill and at Shakespeare Cliff, horizontal curves of radius 6000 m could only just be fitted in place within the Parliamentary limits of deviation.

General aspects of undersea alignment in UK sector

33. By contrast with the underland sector there were few spatial restrictions undersea, and it was therefore possible to seek to optimize the alignment with respect to geology and operation. As there are approximately 60 km of undersea tunnels in the UK sector, the importance of tunnelling conditions to progress and cost is obvious.

34. A reference alignment was prepared before tunnel driving started and was revised as work progressed, taking into account additional geological information, design development at the crossover and pumping stations, and experience gained during driving of the service tunnel itself.

UK coast to the UK undersea crossover

35. Crossovers between the two rail tunnels are provided at the tunnel portals and at the one-third points, to enable a section of the tunnel to be closed for maintenance. The excavation of large caverns undersea was a difficult and potentially more risky undertaking than the TBM drives, and therefore the exact locations of these caverns were chosen in the most favourable geological areas. For this reason, the French crossover was sited at km 44·7, and the mid-point between there and the crossover at the UK portal was at km 27·6. This was close to a zone of possible structural disturbance at km 28 to km 29 associated with an anticline; hence the position of the UK crossover was fixed at km 27·1, which was satisfactory from both geological and operational criteria.

36. The vertical profile between the coast and the crossover was selected in 1987 to comply with the following objectives

(a) the tunnels to be in the lower half of the Chalk Marl

Fig. 6. Variations in geotechnical index parameters with depth

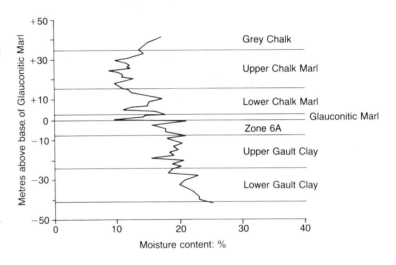

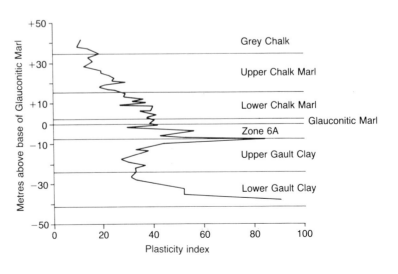

(b) the invert of the crossover cavern to be at the base of the Chalk Marl, so as to maximize the thickness of Chalk marl over the crown of the cavern, and to reduce the rate of seepage infiltration into the cavern

(c) the crossover to be located where the grade is no steeper than 0·9% for operational reasons

(d) the depth of Chalk Marl under the tunnel at the low-point to be maximized, so as to reduce the cost of the sump at the pumping station.

37.   The profile best meeting these objectives has the low-point at km 24·9. Subsequently, the 1988 investigations located zone 6A at the top of the Gault Clay, which enabled the invert of the crossover cavern to be lowered to the base of the Glauconitic Marl, thereby increasing cover over the cavern.

38.   Tunnelling conditions in the undersea service tunnel proved to be good initially but deteriorated just offshore at about km 20·2 and from there until km 28·0. The drive was slowed by overbreak and seepage inflow. Vertical probes sunk behind the TBM in the first 2 km showed that the base of Glauconitic Marl was 2–4 m lower than predicted and it was inferred, therefore, that the cover of Chalk Marl over the tunnel was correspondingly reduced.

39.   Preliminary results from the 1989 investigations at the site of the UK undersea crossover reinforced earlier opinions that the lower Chalk Marl (Fig. 6), was less permeable than the upper part. Calculations highlighted the significance of the thickness of Chalk Marl over the tunnel in reducing seepage inflow (Fig. 7). Therefore, the service tunnel profile was lowered between the TBM position, which was at km 21·7, and the low-point at km 24·9. The revised profile is shown in Fig. 2. Actual tunnelling conditions were worst around km 23 but improved thereafter and then returned to what had been anticipated from km 28·4 through to the junction with the French drive. It is considered that conditions at km 23 might have been worse still had the tunnel not been lowered.

UK crossover to meeting point

40.   The alignment and vertical profile were arranged so that there is only one low-point between the two marine crossovers, at km 35.

41.   The tunnel was located to run parallel to and at least 500 m to the north of the Fosse Dangeard. The tunnel was aligned to maximize the cover of solid rock, to above 22 m, where the tunnel crosses beneath a tributary of the Fosse at km 39.

42.   Additional tight points over this section were a high point in the Gault Clay at km 40 and a low point in the Chalk Marl at km 41. The 1988 investigations showed that there was hardly any faulting, with throws exceeding 0·5 m close to the tunnel, and the improved accuracy of the 1988 geophysical survey allowed the alignment and profile to be optimized from km 38 through to the meeting point. The final alignment is shown in Fig. 2.

Part 3: survey

Initial survey data, grids and datums

43.   Before 1973, the Universal Transverse Mercator (UTM) projection zone 31 was used as a basis for horizontal positional data, the UTM31 grid being added to existing mapping and charts where necessary by the UK Ordnance Survey (OS) and the French Institut Géographique National (IGN).

44.   Since refraction problems preclude precise levelling observations across the Channel, the British and French national levelling networks were connected on the basis of a study of the dynamics of water movement,[2] which identified a sea-slope of about 8 cm between mean sea levels, giving a difference of 0·44 m between Ordnance Datum Newlyn (ODN) and the French datum (IGN69). A Channel Tunnel Height Datum (CTHD) exactly 200 m below ODN was adopted to avoid negative elevations in the undersea tunnels.

45.   The common grid and datum permitted the definition of horizontal and vertical alignments in a unified system, but the inherent distortions between the plane UTM grid and the spheroid would have led to complications in setting out, so it was decided to create a special close-fitting projection, namely the Channel Tunnel Grid (CTG). This is a Cylindrical Orthomorphic Transverse Mercator, with a Central Meridian at 1°30′ East of Greenwich, selected to pass through the centre of the project, creating a plane grid in which locations could be defined and set out with negligible distortion (Fig. 8).

Horizontal control network

46.   Computations in 1974 had combined data from the national triangulations with earlier cross-channel measurements to determine co-ordinates of four UK and six French

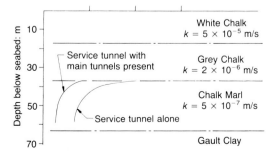

Seepage inflow: (l/s per m run)

Fig. 7. Relationship with seepage inflow rate into service tunnel and altitude of tunnel

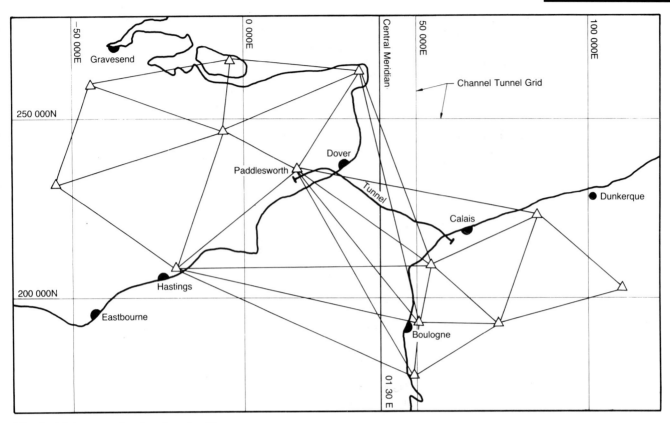

Fig. 8. Main control network and grid

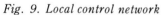

Fig. 9. Local control network

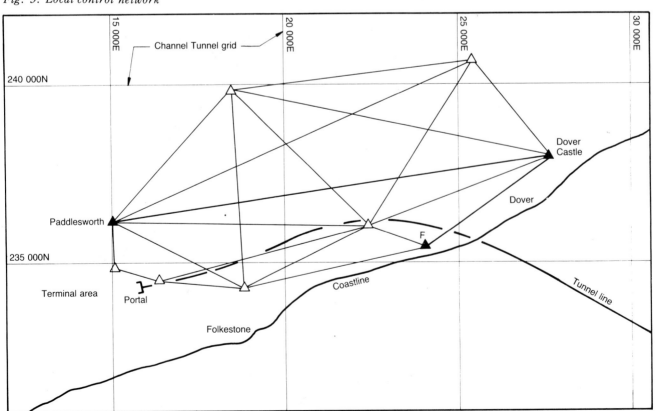

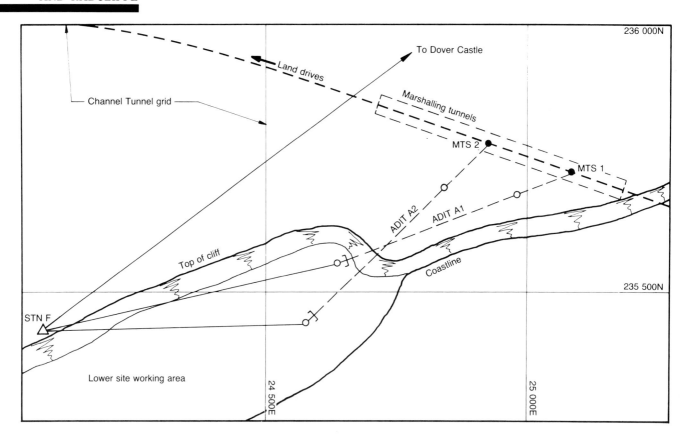

236 000N

To Dover Castle

Land drives

Channel Tunnel grid

Marshalling tunnels

MTS 2

MTS 1

ADIT A2

ADIT A1

Top of cliff

Coastline

235 500N

STN F

Lower site working area

24 500E

25 000E

Fig. 10. Transfer of control to tunnels

primary stations.[3] In 1986, the OS and IGN were commissioned to update the CTG, to provide new control on both sides, and to re-evaluate the datum relationship. The original CTG parameters were retained, but the network of stations by which it was manifested was now considered too sparse, and an enlarged network was computed, incorporating additional secondary observations and other recent data, including astronomic azimuths.

47. In order to provide control points in the works areas, new stations were established to form a local network covering the Terminal and Shakespeare Cliff sites (Fig. 9), and all new observations were included with the previous data in a simultaneous readjustment. To avoid confusion with previous co-ordinate values, the false grid origin of the CTG was shifted and the new network was designated CTG86.

48. Results of a campaign of Global Positioning System (GPS) satellite observations by the OS and IGN in 1987, which were incorporated in a general readjustment computation, improved the overall accuracy, and the revised network, still based on the CTG grid, was designated Reseau Trans-Manche 1987 (RTM87), with overall accuracies estimated as 1 ppm in distance and 0·2 s in direction.

Vertical control

49. In 1986, as no more accurate alternative to the original sea-slope estimation had proved

possible, this was combined with updated mean sea level values to establish a slightly revised datum relationship—IGN69 being 0·442 m below ODN—and site benchmark values were adjusted accordingly.[4]

50. The 1987 satellite results improved the relationship, although still with some uncertainty owing to a lack of precise data on the geoid/spheroid relationship. IGN69 was now established as 0·300 m below ODN, and the CTHD 200 m below ODN was redesignated Nivellement Trans-Manche 1988 (NTM88), all benchmark values being again readjusted.[5]

Transfer of control to tunnels

51. Apart from a short length near the terminal, the UK tunnels were constructed from the coast at Shakespeare Cliff, three headings being driven about 8 km inland and three between 19 km and 22 km seaward to meet the French drives. From the coastal work area west of the tunnel line and 16 m above sea level, two inclined adits descend to the marshalling tunnels, where all drives began. The adits offered a suitable route for alignment transfer, as both portals were visible from a station (F) of the Control Network on the cliff-top, about 550 m to the west (Fig. 10). This made possible a very direct transfer of azimuth from the primary station at Paddlesworth (P) via Dover Castle (DC) and (F) to a loop traverse through the adits, incorporating the baseline MTS1–

MTS2 from which tunnel alignment could begin in both directions.

52. From stations of the local network, precise levelling was carried down the adits, establishing new benchmarks in the marshalling tunnels. The subsequent completion of an access shaft from the upper site permitted an independent check to be made, using vertical EDM measurements.

Maintenance of alignment

53. The maintenance of alignment comprised two stages of survey, namely primary control and secondary control.

Primary control

54. The primary horizontal control is, in effect, an underground extension of the main surface network, and its stations require to be as stable as possible and to provide repeatable forced centring for instruments. Pre-formed fixing holes in the concrete linings permitted the design of bracket stations, mounted with anchor-bolts and incorporating instrument tables to accommodate gyrotheodolites as well as the normal survey instruments and accessories (Fig. 11).

55. The configuration of the primary network in the service tunnel was influenced by the layout of the services required for the driving process. This provided space for personnel safety refuges on one side only, and ini-

tially it was decided that the network would have to take the form of a single traverse along that side, using the refuges as observation platforms. Such a configuration lacked rigidity and was vulnerable to lateral refraction effects, but was sufficient to allow driving to begin. Further study of the clearances available on the opposite side showed that it was possible to design a bracket which would remain just clear of trains (Fig. 11), and the installation of these right-hand stations permitted the observation of closed polygons and zigzag traverses which showed significant azimuth differences from the single-side route.

56. Analysis of the observations and comparison with azimuths from the newly acquired gyrotheodolite showed that the sightlines of the single-side traverse were being curved in one sense only by temperature gradients across the line, while those of the zigzags had a reverse curve form.

57. Therefore, the observed angles of the single-side traverse tended to be systematically too large, whereas those of the zigzag were alternately too large and too small by similar amounts, thus tending to compensate in overall azimuth. The principle of symmetrical zigzag traversing was therefore adopted for all primary control, and gyrotheodolite azimuths, whose accuracy is independent of distance traversed, were incorporated at frequent intervals.

58. The maintenance of vertical control was

Fig. 11. Survey stations in service tunnel

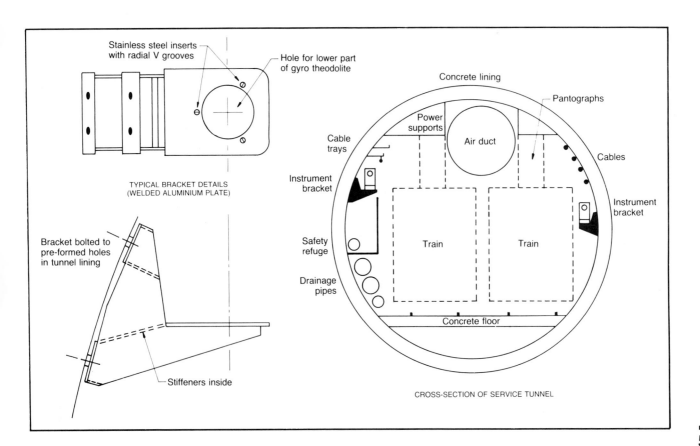

relatively straightforward, with permanent stainless steel benchmark bolts being installed at 75 m intervals and their elevations established by closed loops of precise levelling using parallel-plate micrometer levels and invar staffs.

Secondary control

59.   The secondary horizontal control consisted of forward traversing from the primary stations to control the steering of the tunnel boring machines (TBMs). The congestion of the equipment forming the TBM support train, over 250 m long, limited the secondary control to single-side traversing in a confined space. Temporary bracket stations were installed alongside the support train and surveyed in. These provided mountings for the laser and its associated 'gates' defining the reference line for the ZED guidance system incorporated into the TBM. The laser beam was set up to a predetermined location and alignment so as to be received on a screen mounted on the cutting head, and details of its position were loaded into an on-board computer which established the location of the head by reference to the laser and compared it with the designed tunnel alignment (DTA) previously stored. A digital display of discrepancies between designed and actual TBM position permitted the driver to make continual alignment corrections by steering to 'zero'.

Tunnel junctioning and survey closures

60.   The first UK breakthrough was of the 8 km land service tunnel. A preliminary alignment check was made via a vertical borehole drilled from the surface to intersect the tunnel line ahead of the TBM and about 200 m back from the break-out point. A precise optical plummet was set up above the borehole, its co-ordinates established, and a laser attached. Once the TBM had cut through the bottom of the borehole, the vertical laser beam was visible behind the cutting head and its position was surveyed in relation to the tunnel control stations. This preliminary closure showed that no alignment correction was required, and the TBM later broke out accurately into its dismantling chamber. The subsequent control survey closure gave discrepancies of only 4 mm laterally and 15 mm vertically. Later, the land running tunnel traverses were linked to the service tunnel control at the most forward available cross-passage before break-outs, which were achieved with similar accuracies.

61.   To establish a preliminary survey closure for the marine service tunnel, the UK TBM was halted and a horizontal borehole was drilled forwards about 100 m while the French TBM continued driving. The borehole was surveyed using a Reflex MAXIBOR instrument and the French team was notified of the position at which they could expect to find it. The French TBM later cut through the end of the hole and its position in relation to their control stations was surveyed. The comparison of the two sets of co-ordinates for a common point on the borehole permitted the definition of a preliminary corrective line for a further 50 m drive of the UK TBM, while the French TBM was being dismantled from within its casing.

62.   The UK machine was then turned aside on a tight curve to arrive at a pre-arranged burial position, and a small adit was hand excavated from the side of the diverted UK tunnel into the face of the French drive. Following the formal breakthrough on 1 December 1990, the two primary control networks were connected by way of the adit, each survey team working independently before comparing results. The final agreed closure differences were: chainage: 75 mm; lateral offset: 350 mm; elevation: 60 mm.

63.   The alignment over the final 50 m (hand excavated and lined) was readjusted in the light of the closure results to produce a smooth junctioning of the two drives.

64.   For the marine running tunnels, the junctioning was achieved by diverting each UK TBM downwards to be buried below floor level and driving the French TBM through to meet the last UK permanent lining. The control surveys were linked by way of the service tunnel at the closest available cross-passages, and corrective alignments were applied where necessary over the relatively short drive distances that remained, both French TBMs achieving accurate breakthroughs.

References
1. CRIGHTON G. S. *et al.* Supplementary site investigations for the Channel Tunnel 1986–7. *Proc. 5th Int. Symp. on Tunnelling,* London, 1988. IMM, London, 1988, 55–68.
2. CARTWRIGHT D. E. and CREASE J. A comparison of the geodetic reference levels of England and France by means of sea surface. *Proc. R. Soc.,* Series A, 1963, **273**, No. 1355, June.
3. ANON. *Adjustment of primary control for Channel Tunnel survey.* Ordnance Survey, 1974.
4. CALVERT C. *Channel Tunnel grid prepared for Transmanche-Link.* Ordnance Survey, 1986, Dec.
5. ANON. *Reseau du Tunnel sous la Manche.* IGN France International, 1987, Dec.

Machine-driven tunnels

P. J. Chapman, V. H. Wrightam, G. Ince and R. P. Lewis

Proc. Instn Civ. Engrs, Civ. Engng, Channel Tunnel Part 1: Tunnels 1992, 55–86

Paper 9964

■ **A major element of the UK underground construction works was the requirement for 84 km of tunnels to be constructed by tunnel boring machines (TBMs). On the UK side, the programme required six TBMs, each of which was specifically designed and built for high-speed tunnelling, and continuous around the clock working. The Paper describes the design, manufacture and basic technical data for the TBMs, together with the installation and commissioning underground. The operation of the TBMs is also described, as are the modifications which were made to improve their efficiency. TBM and immediate support crew manning levels, rates of progress and detailed cycle times are covered, together with methods of overcoming specific problems encountered in the individual drives.**

Facts determining choice of tunnelling method

As far back as the 1880s, when Colonel Beaumont's tunnel boring machine was used for the first Channel Tunnel attempt, it was believed by engineers that a tunnel boring machine (TBM) was the economic solution. In the 1980s, the same was true, but the choice had narrowed to that between different types of TBM; a choice governed by two main factors: the geology and the programme.

Geology

2. *Site investigations* Geotechnical investigations have been carried out across the Channel since 1875, and well over 100 boreholes have been sunk. Most recently, the 1980–87 investigation included a geophysical survey. All the available data indicated that the tunnelling on the UK side could be confined to the Chalk Marl.[1] Chalk Marl is a clayey carbonate mudstone, found in layers which can range from moderately strong carbonate beds to a weak clay-rich material. The Chalk Marl was believed to be essentially impermeable, with any water restricted to discontinuities, such as faults or fissures. It was considered to be an ideal tunnelling medium, as it was easy to excavate and possessed a 'suitable' stand-up time for TBMs. A general geological cross-section is shown in Fig. 1.

3. *Probing philosophy and inundation risk* Despite the extent of the site investigation, this still gave boreholes at only about 1 km centres. There was, therefore, an element of risk which was to be dealt with by using the service tunnel as a pilot tunnel and probing ahead of its face.

4. In early 1988, a specific study was made of the risk of inundation and of the requirements for probing. In parallel with this study, trials were carried out to prove both the capability of drilling equipment, the results obtainable from various 'down-the-hole' logging tools, and the potential of seismic tomography.

5. The study identified six possible sources of geological hazards which could be located by probing.

(a) Water-bearing fissures. It was recognized that water-bearing fissures could go undetected in the hand-mined sections, as well as in the TBM drives. However, the greatest risk of sudden inundation was in the TBM drives.

(b) Site investigation boreholes. Boreholes had been drilled in the Channel over a period of many years, and locations of the early ones were not precisely known. The study showed that five boreholes, drilled and grouted in 1972–73, were within 70 m of a running tunnel, the closest being 26 m, although the locations of these holes were known only to within ±10 m.

(c) Deep weathering.

(d) Proximity to the Grey Chalk.

(e) Proximity to Gault Clay. Aspects (c)–(e) were more related to a concern for the stability of excavation than to inundation, although in the case of (d) to (e), they could result in water-bearing fissures.

(f) Proximity to the Fosse Dangeard Fault.[1,2] This was considered to warrant very close examination, and much use was made of knowledge of the geology acquired during the advance of the service tunnel.

6. The study concluded that the risk of inundation was very small, provided that adequate probing was carried out. In this respect, therefore, it was decided that during the service tunnel drive, two 100 m long holes should be destructively drilled at 4° to the longitudinal axis and at 3° elevation, at intervals of 75–80 m. In addition, downward vertical holes should be drilled at 75–500 m intervals, depending on the proximity of the Gault Clay, and 35 m long sideways horizontal holes should be drilled at intervals of 37·5–187·5 m. Information provided by 'down-the-hole' logs and seismic tomography were not considered to be of sufficient value to warrant the considerable

P. J. Chapman, Tunnel Construction Director, UK Construction, Transmanche-Link JV

V. H. Wrightam, Senior Agent, Tunnel Construction UK Construction, Transmanche-Link JV

G. Ince, Engineering Director, James Howden Ltd

R. P. Lewis, Director of Tunnelling, Robbins Markham JV

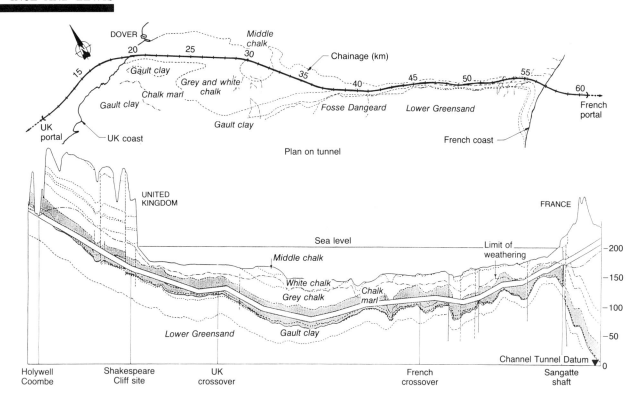

*Fig. 1. Layout and
geological section*

disruption to the tunnel drive which they would
cause.

Programme

7.   The project programme called for all
bored tunnelling to be carried out in a $3\frac{1}{2}$ year
period, between December 1987 and June 1991.
This requirement dictated the need for six
TBMs capable of rates of advance of 5 m per
hour for the service tunnels and 4 m per hour
for the running tunnels.

Type of TBM selected

8.   The selection of the TBM had to be considered as one part of a whole system, and the
programme and the geology also influenced the
choice of tunnel lining and back-up system.

9.   On this basis, it was decided that the
most suitable tunnelling system would be an
open, full-face TBM, used in conjuction with
unbolted, expanded precast concrete linings.
Geotechnical studies indicated that 95% of the
drives would be suitable for this type of lining.
The spoil handling would be carried out by conveyors that would load the chalk marl from the
cutting head into 14 m³ side tipping skips.
These would tip into a bunker storage system
at the Shakespeare marshalling area, which
would deliver the material by means of a conveyor system to the surface, where it would be
used as landfill behind a new seawall.

10.   To meet the very high sustained rates of
progress required by the programme, the TBMs
had to be of a type which would allow the erection of linings to proceed concurrently with

face excavation. This is called for a telescopic
TBM, in which the front body contained the
excavating cutterhead and the rear body contained grippers. The front body had to be
capable of being pushed forward 1·5 m, using
the gripper unit as a reaction point, while a
1·5 m long ring was being built concurrently at
the rear of the TBM, about 13 m behind the
face.

Procurement of TBMs

11.   *Numbers of TBMs required*   After the
type of TBM required had been selected, the
next decision was to determine the number and
direction of drives and, hence, the number of
machines required, and how they would be used
to achieve the programme.

12.   It was concluded that the programme
could be achieved only by dividing the tunnel
drives into two groups, namely land and
marine. Furthermore, the degree of concurrency
demanded by the programme meant that up to
six TBMs would be required.

13.   For the marine tunnels the choice was
clear. They were to be driven from below the
coastal site at Shakespeare Cliff to the junction
point with the French TBMs, about 22 km offshore, and three TBMs would be required: i.e.
one for the service tunnel and one each for the
two running tunnels.

14.   For the land drives, the choice was more
complex. They could be driven uphill from
Shakespeare Cliff to the Holywell cut-and-cover
section, or downhill in the reverse direction. If
driven uphill, they could be pushed across the

Holywell cut-and-cover and be used for the construction of the 500 m long Castle Hill tunnels.

15.   After much discussion on programme, logistics, geology, distribution of tunnel spoil, environmental and other considerations, it was decided to drive uphill to Holywell but not to use the machines for the Castle Hill tunnels. In order to remove an unacceptable concentration of activity at Shakespeare Cliff, the programme was arranged to require only five TBMs to operate at any one time, the land service tunnel being programmed to finish before the land running tunnel south drive started (see Fig. 2).

16.   *TBM procurement*   The first enquiry for TBMs was issued to manufacturers in February 1986. TML issued a performance specification to tenderers which, *inter alia*, specified the following criteria: open face, full face TBMs; instantaneous rate of penetration; cycle times for one ring (excavation and build); size and type of lining; length of drive; minimum maintenance requirements; an ability to contain the possibility of inundation if the machine encountered an old borehole (marine drives); type of ground to be traversed; life of 20 000 actual operating hours; cutterhead to be dual speed and bi-directional; cutterhead to be suitable for discs or picks; ability to probe ahead of the TBM.

17.   It had been agreed with Eurotunnel that the order for all four running tunnel TBMs would not be placed with one supplier.

18.   The service tunnel TBMs were ordered from James Howden & Co of Glasgow in November 1986. The marine running tunnel TBMs were ordered from Robbins Markham Joint Venture in August 1987, and the land

running tunnel TBMs from Howdens in December 1987.

Choice of lining design

19.   In parallel with the TBM design and procurement, a design for tunnel lining was developed.[3,4]

20.   The geology indicated that an expanded lining would be suitable and the programme indicated that the speed of erection offered by an expanded lining was the correct choice.

21.   The method of expansion chosen was by use of a wedge-shaped key, supplied in three different sizes to cater for variations in cut diameter. Each segment was to have four 20 mm thick pads on its back to provide initial contact with the excavated profile, leaving a 20 mm void after expansion of the ring. The annular void was then to be filled with grout to provide full contact with the ground. A summary of ring types is shown in Table 1.

22.   Following early experience with the service tunnel linings, it was decided that, to save costs and drive time, a hybrid lining of precast concrete (PCC) and spheroidal graphite cast iron (SGI) segments should be designed, to allow the rapid construction of openings for both cross-passages and piston relief ducts. The resulting SGI elements were pre-assembled into panels 1500 mm wide, which matched exactly the external dimensions of the PCC segments and, by use of adaptors, could be handled by the TBM erectors in the same manner. Opening sets could, therefore, be erected in almost the same time as a standard ring. The design also saved time in opening construction, compared with the more conventional design of bolted beams and jamb plates.

Fig. 2. Tunnel plan showing programme at March 1990

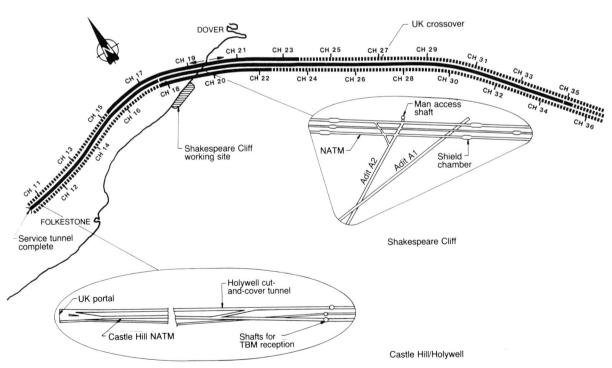

For other areas, i.e. in adverse ground conditions, conventional bolted and gasketted SGI rings were provided. These were cast as 750 mm wide segments and pre-assembled on the surface into 1500 m wide rings.

Details of TBMs

Service tunnel TBMs

23.   Both land and marine drive machines were designed and built by James Howden & Co. They were designed to the same specification, in order to use the same major components and to allow interchangeability of spares; although, owing to different lining thicknesses, their cut bores were 5·38 m for the marine tunnel and 5·76 m for the land tunnel. The major common components included the main bearing of three-row roller design with integral drive gear, four two-speed reversible electric 190 kW drive motors, and clutches, gearboxes, shove rams and power packs.

24.   The cutterhead consisted of four radial arms dressed with a double array of 68 rocking picks for the marine service tunnel, and 76 rocking picks for the land service tunnel, together with ten profile gauge picks. If ground conditions required, these could be replaced by disc cutters.

25.   A drum type design was used for the cutterhead, with an outer conical rotor and inner fixed drum, into which the spoil conveyor protruded. The cut spoil was lifted by loading blades and dropped through a central aperture, into a hopper, over the conveyor. The front section of the conveyor could be withdrawn from the head in the event of inundation (see also §§ 67–69).

26.   The specified cycle time was 18 min, with ring building taking place simultaneously

Table 1. Summary of ring types used in UK tunnels

Ring type	Nominal inner dia.: m	Thickness (excl. pad): mm	Width: m	Weight per ring: t	Tunnel
PC concrete					
S27 A, B, C*	4·8	270	1·5	15·98	Marine service
S41 A, B, C*	4·8	410	1·5	25·41	Land service
R27 A	7·78	270	1·5	25·37	} Marine running
R36 A, B, C*	7·6	360	1·5	33·15	
R54 A, B, C*	7·6	540	1·5	51·08	Land running
Heavy duty SGI					
SB1	4·8	410	0·75	2·12	} Marine service and
SB2	4·8	140	0·75	2·72	ancillary structures
SB3	4·8	150	0·75	4·74	Land service and ancillary structures
RB1	7·6	150	0·75	4·16	} Marine running
RB2	7·6	150	0·75	5·88	
RB3	7·6	175	0·75	10·21	Land running
SGI opening sets					
SZ2, 3	4·8	120	6·0	19·00	Marine service
SZ7	4·8	140	6·0	33·20	Land service
Hybrid cross-passage opening sets (PCG/SGI) SGI elements only					
R27 X	7·6	270	6·0	18·52	} Marine running
R36 X	7·6	360	6·0	24·12	
R54 X	7·6	540	6·0	35·32	Land running
Hybrid piston duct opening sets SGI elements only					
R27 P	7·6	270	3·0	6·90	} Marine running
R27 P/1	7·6	270	6·0	13·80	
R36 P	7·6	360	3·0	9·10	
R36 P/1	7·6	360	6·0	18·20	
R54 P	7·6	540	3·0	14·00	} Land running
R54 P/1	7·6	540	6·0	28·00	

* Variations in code letter where dimensions are the same indicate variation in internal reinforcement for different ground loading.

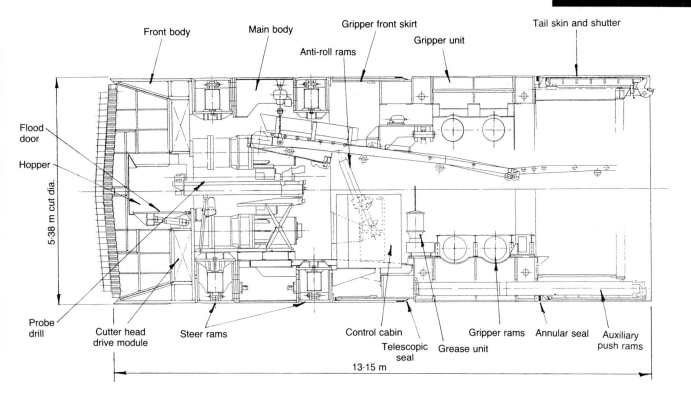

Front body · Main body · Gripper front skirt · Anti-roll rams · Gripper unit · Tail skin and shutter

Flood door · Hopper · 5·38 m cut dia. · Probe drill · Cutter head drive module · Steer rams · Control cabin · Telescopic seal · Grease unit · Gripper rams · Annular seal · Auxiliary push rams · 13·15 m

Fig. 3. UK service tunnel TBM

with excavation. The overall length of the machine was 13 m, comprising a three-section telescopic body, encompassing cutter head, main body and gripper, together with the shove and build ram assemblies. Steerage was effected by the shove rams and by pairs of front and rear hydraulic steer shoes on two axes. To provide reaction against the cutterhead torque and to correct roll, diagonally opposed hydraulic cylinders were fitted between the rear of the main body and the front of the gripper section (see Fig. 3). A 40 m bridge-beam carrying segment handling and erector system was towed behind the TBM.

27. A ZED 260 laser-activated guidance system with instantaneous visual call-up of the TBMs' three-dimensional position and attitude was contained in the main control module. Loading of information into the inboard computer could be made by way of a remote station further back on the TBM sledges.

28. Ring building took place directly behind the gripper section where the lower three segments were built using a rack-mounted slewing boom crane. A 180° upper erector was used to build the three upper plates and key. Segments were supplied from flatbed wagons, by way of a second rack-mounted crane, either to the lower crane erector or into an upper magazine.

29. Segments were attached to the erector by a quick-release pin which was located in a cast iron pot, cast into each segment. The pin mechanism was activated manually in the case of the lower cranes, and hydraulically in the

case of the upper erector and magazine loader.

30. PCC planks, spanning between haunches formed in the segments, were installed 20 rings back from the face to carry the 900 gauge tracks. These planks later form part of the permanent service tunnel deck.

31. Gantries spanning the single off-centre track access to the segment unloading area completed the 180 m long TBM. The gantries housed the spoil conveyor power units, services, grouting equipment, and ventilation cassette-handling facilities. The length of the gantries enabled a 102 m long train to be accommodated at any one time, thus allowing a two-ring (3 m) cycle of excavation and ring erection to be carried out per train visit. The spoil conveyor was of telescopic design, moved longitudinally by a rack to allow the spoil to be loaded evenly into a stationary train.

32. A Schwing grout pump, bulk cement silo and powder transporter tanks were supplied originally, and worked well in the relatively dry conditions of the land service tunnel. However, this system was replaced on the marine service tunnel, by vertical pan mixers, mono pumps and bagged materials.

33. Ventilation was provided to the rear of the TBM by means of a single 1·2 m dia. flexible duct, fed from cassettes, each containing 150 m of duct. Fresh air was carried forward to the build area in twin 300 mm dia. ducts. Korfman dry type de-duster units were also incorporated to purify recirculating air.

34. Power to the marine service tunnel TBM

was supplied by a twin 11 KVA ring main circuit system, fed by way of sub-stations situated in the cross-passages, excavated 1 km behind the tunnel construction zone. This allowed the 11 KVA power supply to be extended without interruption to the drive.

35. The marine TBM was equipped with telescopic drainage pipe connection equipment, and ingress and cooling water were pumped to waste by way of a series of temporary pump stations, also situated in cross-passages.

36. The land service tunnel TBM was served by a single 11 KVA power supply. Cooling water and small quantities of ingress water were pumped and piped back down the grade to Shakespeare Cliff marshalling area, where they were pumped to the surface for cleaning, before recirculation and re-use.

Marine running tunnel TBMs

37. The two marine running tunnel TBMs were built by Robbins Markham JV to identical specifications and provided a 8·36 m cut diameter. The TBM consisted of a forward cutterhead unit, linked telescopically to the rear gripper section by a series of angled propel rams. The rear of the gripper unit was fitted with a series of shove rams to assist ring building and gripper advance. This system allowed

simultaneous excavation and ring building (see Fig. 4), and was designed to achieve a cycle time of 22 min.

38. The cutterhead consisted of a total of 195 picks mounted on eight radial arms. There was a facility to replace these picks with 60 disc cutters or to use a combination of picks and discs. The cutterhead was driven by 12 two-speed reversible electric motors, giving a head rotation speed of either 1·67 rev/min or 3·33 rev/min.

39. The spoil was handled by a system of three belt conveyors, carried within the back-up gantries. The primary conveyor was capable of being withdrawn from the head for maintenance and inundation protection (see also §§ 67–69). The primary conveyor fed, by way of a fixed secondary conveyor, on to a shuttle conveyor with a belt reversal system, capable of filling a train for one complete ring (1·5 m advance). The back-up gantries also carried main transformers, hydraulic power packs, emergency battery power, de-duster fans, compressors, grouting and other ancillary equipment. At the rear of the gantries were cable and service water reeling drums, together with a 2·0 m dia. ventilation cassette. These facilities allowed continuous operation of the machine.

Fig. 4. UK marine running tunnel TBM

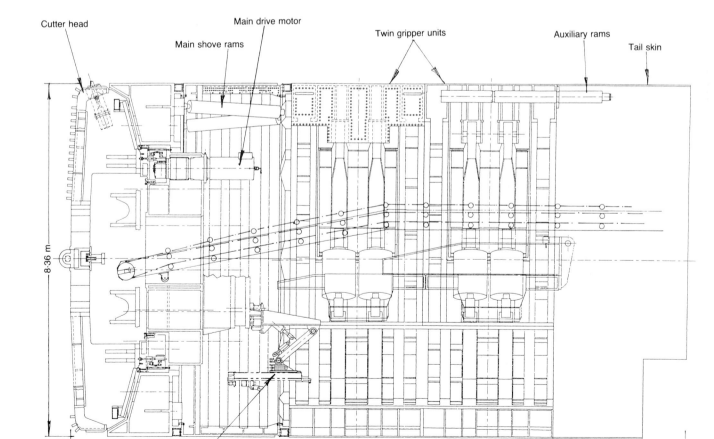

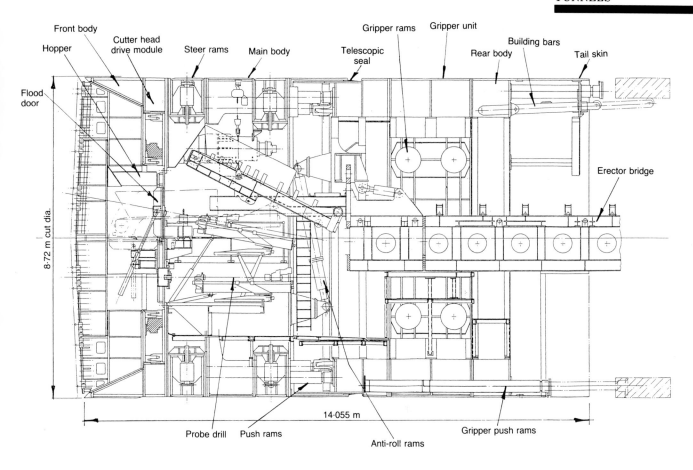

Flood door

Hopper — Front body — Cutter head drive module — Steer rams — Main body — Telescopic seal — Gripper rams — Gripper unit — Building bars — Rear body — Tail skin

8·72 m cut dia.

Erector bridge

Probe drill — Push rams — Anti-roll rams — 14·055 m — Gripper push rams

Fig. 5. UK land
running tunnel TBM

40. The segment handling comprised a lower system for the invert and two haunch segments, and an upper system for the remaining segments. The segments were unloaded using electrically operated overhead travelling cranes which fed on to upper and lower slat conveyors. The slat conveyors were capable of holding two complete rings of segments.

41. The segments were fitted with a cast iron lifting pot. The erector head was fitted with a hydraulically operated, tapered, oval pin. The pin was inserted into the lifting pot, and rotated through 90° to lock and unlock.

42. Segment erection was carried out under a flexible hood which provided additional protection and support in the build area. This consisted of steel fingers between which stainless steel plates were fitted, spanning from the tailskin to the last ring.

43. The lower three segments were erected using two lower erector cranes, the invert segment always being built one ring in advance, within the protection of the tailskin. The remaining segments were fed by the upper slat conveyor to the two upper erectors. The ring was completed and held using build bars on the gantries. The gripper section and hood was then advanced 1·5 m before the key was fully inserted to expand the ring, thus reducing the risk of trapping the flexible hood.

44. The overall length of the TBM and gant-

ries was 250 m, with a total weight of 1500 t. A double rail track 900 gauge ran under the gantries. This was supported on pairs of precast concrete sleeper blocks, pinned to the segments and stabilized with track ballast. While one track was occupied by a train being loaded with spoil and having segments unloaded, the other track was free to receive the next train. Each train carried a full ring and consisted of five flat cars with segments, cement and other materials, and eleven 14 m^3 muck cars.

45. From the TBM cab, the operator controlled the excavation and alignment of the TBM using a ZED 260 guidance system. The cab was equipped with CCTV monitors that showed conveyor operations, and with VDUs that displayed the operational status of the TBM.

Land running tunnel TBMs

46. Both machines were designed and built by James Howden & Co. to identical specifications in order to maintain interchangeability of components, including electric drive motors, clutch and gearbox assemblies, shove rams and hydraulic pumps. To meet the specification of one complete cycle in 22 min, a telescopic TBM was again selected, consisting of a cutterhead/ front section, linked telescopically to a gripper

Table 2. Tunnel boring machine data

Description	Service tunnel Howdens*	Land running tunnels Howdens	Marine running tunnels Robbins Markham
Cut diameter MS	5·38 m		
LS	5·76 m	8·72 m	8·36 m
Design advance rate	5 m/hour	4 m/h	4 m/h
Machine weight (approx.)	625 t	1757 t	1350 t
Machine length	13 m	15 m	15 m
Cutters MS	68 rocking picks	268 rocking picks	276 rocking picks
(alternatively) LS	76 rocking picks	(70 discs)	(58 discs)
	(40 discs)		
Cutterhead torque	1424 kNm (H)	2305 N m (H)	5250 kNm
	2137 kNm (L)	4807 N m (L)	
Cutterhead speed	4·5(H) 3·0(L) rev/min	2·8(H) 1·9(L) rev/min	2·2(L) 3·3(H) rev/min
Cutterhead drive	4 × 190 kW	6 × 190 kW	12 × 110 kW water cooled
Number and stroke of advance rams	16 × 1600 mm	24 × 1600 mm	20 × 1500 mm
Installed thrust to cutterheads	2072 t	3120 t	4220 t
Number and stroke of auxiliary thrust rams	7 × 2100 mm	9 × 2100 mm	16 × 2000 mm
			1 × 2000 mm key ram
Installed thrust of auxiliary rams	130 t	1170 t	1600 t
No. of gripper shoes	2	2	4
Ground pressure from gripper (normal)	21·45 bar	21·45 bar	18·5 bar

* Name of manufacturer.

section by shove rams.

47. The overall length was 15 m and the weight 825 t. The front section of the TBM carried the cutterhead and the steering shoes, and the gripper section housed the gripper pads, gripper body push rams and auxiliary rams (see Fig. 5).

48. The cutterhead had eight radial arms, dressed with a quadruple helical array of 214 pick cutters, with 54 gauge picks and 12 profile picks. If ground conditions required, these could be replaced by disc cutters.

49. A drum type design was used for the cutterhead, with an outer conical rotor and an inner fixed drum, into which the conveyor protruded. The cut spoil was lifted by loading blades and dropped through a central aperture into the hopper over the conveyor.

50. A single continuous belt conveyor transported the spoil a distance of 200 m and by way of a travelling side-discharge conveyor (tripper), into trains of seven 14 m³ side tipping skips. Each was sufficient for half a ring advance (0·75 m) and could be loaded on either of the twin rail tracks on the lower deck of the TBM back-up trolleys.

51. Tunnel alignment was monitored from the TBM cabin, by a ZED 260 laser guidance system. Steering was achieved by manual selection of the push rams housed in the telescopic section. Additional steering bias could be effected by pairs of front and rear hydraulic steer shoes located in the front body and acting on each of the two axes. Torque reaction from the cutterhead was taken into the gripper through diagonally opposed hydraulic cylinders fitted between the front body and the gripper, which could be used to correct roll.

52. Two independently powered fully rotational 360° erectors were provided to allow concurrent segment building on both sides of the tunnel.

53. Following each 1·5 m advance of the TBM gripper section, and during the excavation of the next 1·5 m advance, a ring of segments was assembled by the two erectors which were mounted on slides and travelled longitudinally on a 14 m long erector bridge. The erector bridge spanned the build area and was supported between the gripper unit and the front platform. The TBM control cabin was mounted on top of the front platform.

54. The segments were erected in a two-ring build configuration, the three lower segments of the leading ring being built after the five upper segments and key of the previous ring had been erected and expanded.

55. Located behind the erector bridge and front platform was a 56 m long crane bridge, supported by two hydraulically adjustable portal wagons. The front section of the crane bridge was designed as a cantilever, 16 m long,

and was therefore structurally independent of the front platform, apart from the towing bars.

56. The segments were supplied on flat cars, propelled at the front of the material train. They were unloaded by two electric travelling cranes, one mounted on each side of the crane bridge. The cranes carried the segments forward and placed them on the segment magazine in the invert. The segment magazine was of rack and pawl design and fed the segments forward into the build area where they were picked up by the erectors.

57. The erectors worked in conjunction during ring building, each lifting and placing alternate segments on either side of the tunnel. If one erector broke down, the remaining erector could be used independently and was capable of building a full ring.

58. The erectors engaged the segments using two 115 mm dia., hydraulically operated pins, housed in the erector head. The pins were inserted at a converging angle into similarly converging lifting pots, cast into the segments. The segments were locked on to the converging pins by a wedge action, using smaller hydraulic cylinders to clamp and secure.

59. The erectors were controlled from three control panels, located strategically on the erector platform. A Castell key system was employed to prevent unco-ordinated operation of the separate panels. The control panels also contained a safety logic which required the selection of more than one button to initiate certain critical operations, and thus prevented

the accidental release of a segment during lifting.

60. The erector control panels were used to operate the building rams and key ram, and were also used to operate the hydraulic building bars, which provided temporary support to the upper segments, before insertion of the key. The erector control panels also had master control over the gripper rams, to prevent inadvertent release during ring building. On completion of the ring build, this control was handed back from the erector driver to the TBM driver.

61. The hydraulic power packs, ventilation equipment, cooling system, grouting equipment and electrical switchgear were mounted along the top deck of the crane bridge and portal wagons.

62. At the lower level, the material and spoil wagon trains ran on twin tracks, fixed on the continuous closed deck of the back-up portals. All were towed forward by the advancing TBM. The trains gained access to this deck by way of a 25 m long ramped section of track, towed at the rear of the machine, and were able to travel forward on the lower deck to within 12 m of the segment magazine.

63. Underneath the cantilever section of the crane bridge, and between the front of the lower deck and the rear of the segment magazine, was an open invert area which allowed access for preliminary cleaning of the segments and provided a rail-up area for the installation of the twin track, 900 mm gauge, tunnel construction

Table 3. TBM back-up and ancillary equipment

Description	Service tunnel Howdens*	Land running tunnels Howdens	Marine running tunnels Robbins Markham
Machine length (overall)	180 m	280 m	250 m
No. of back-up sledges	17	29	14
No. of conveyors	1	1	2 fixed, 1 shuttle
Conveyor capacity	750 t/h	1400 t/h	1500 t/h
No. of segment sectors	1 upper and 1 lower	2 × 360°	2
Segment delivery system	Upper magazine Lower crane	1 lower magazine	1 lower conveyor
Fire fighting system	1 water curtain AFFF foam systems Auto Halon systems Hand appliances	2 water curtains AFFF foam systems Hand appliances	1 water curtain AFFF foam systems Hand appliances
Electrical supply	11 kVA/415 V	11 kVA/415 V	11 kVA/415 V
Total installed power	1·2 mW	2 mW	2·3 mW
Emergency battery	900 ah	—	900 ah
Water pumps	2 × 100 l/s inrush 1 × 40 l/s discharge		2 × 150 l/s inrush 1 × 40 l/s circulation 1 × 50 l/s discharge
Ventilation system	250 m³/min incoming 200 m³/min extracted	400 m³/min	300 m³/min incoming 250 m³/min extracted
Probe drills	2 Diamec 260	1 Diamec 260	2 Diamec 260

* Name of manufacturer.

Fig. 6. Running tunnel erection chamber

railway. The twin track was supported on pairs of concrete sleeper blocks which were handled and placed in position by 1 t ancillary cranes attached to the segment cranes. The twin 900 mm gauge tracks were installed ahead of the TBM back-up portals and, therefore, were able to provide support and guidance to the wheel-mounted portals.

64. The two outer rails were used to provide this support (except under the heavily loaded crane bridge podests where hydraulically damped rollers distributed the load over all four rails). The sleeper blocks and track were subsequently stabilized with ballast, after emerging at the rear of the TBM.

65. The overall length of the land running tunnel TBMs and back-up equipment was 280 m, with a total weight of 1700 t. Howden employed ROWA of Switzerland as their specialist back-up system designer.

66. A comparison of the data of the three types of TBM purchased for the project can be seen in Tables 2 and 3.

Inundation protection

67. In specifying the TBMs, it was decided that they should be capable of withstanding inundation arising from the encountering of an unsealed borehole or open fissure within the TBM face, or inundation caused by passing close to an unsealed borehole which could burst into the excavation at the tail of the TBM.

68. *Marine service tunnel* Protection against inundation was as follows.

(a) *Forward probing.* The TBM was equipped with two Diamec 260 probe drills, complete with monitoring devices to measure thrust, torque, water flows and penetration rates, and probing was carried out to detect any adverse geological conditions which may give rise to inundation.

(b) *Inundation at the face.* In the event of inundation, the TBM was protected from the ingress of water through the face by a hydraulically operated door, mounted in the central conveyor aperture. Retraction of the front end of the spoil conveyor from the aperture and closure of the door was controlled by the TBM driver.

(c) *Inundation at the rear.* A sectionalized annular sealing blade was contained in the rear of the gripper. This blade could be hydraulically expanded into the surrounding ground to prevent face inundation from passing along the outside of the TBM to the build area.

69. *Marine running tunnels* Protection against inundation was as follows.

(a) *Inundation at the face.* On activation of an emergency 'inundation button', the central conveyor access hole was closed and sealed following withdrawal of the hopper and primary conveyor unit. This withdrawal was hydraulically operated and took only 15 s. Simultaneously, peripheral knife seals expanded to block the annular gap around the TBM body. The gripper pressures were enhanced to enable them to withstand the 6600 t force exerted by the maximum hydrostatic head of 12 bar acting on a sealed TBM. Fortunately, these measures were never used, but they were always available.

(b) *Inundation at the rear.* At design and manufacturing stage, it was attempted to install shutters which were intended to provide protection against inflows behind the machine. However, when the tunnelling conditions deteriorated, ground support by trailing fingers was more important. Installation of trailing fingers prevented the shutters from being used and their function was abandoned.

(c) In the event of inundation, pumps capable of handling 900 l/min could be started and

an evacuation alarm sounded.
(d) A standby battery and DC generator system were installed which could operate pumps and safety equipment in case of mains failure.

TBM fire protection

70. All of the TBMs were fitted with fire protection systems, as follows.

71. *Service tunnel TBMs* In the service tunnel TBMs, automatic Halon gas systems, activated by heat sensors, were installed at the compressors and electric transformers. Manually activated foam systems were installed for all other high-risk areas. Heat sensors activated both visual and audible alarms to alert the TBM personnel, who assessed the severity of the fire before activating the system.

72. At the rear of the TBM back-up system, a manually operated water curtain was fitted to contain smoke within the TBM area.

73. *Land running tunnel TBMs* Except on the 11 kV switch panels, Halon systems were not fitted to the land running tunnel TBMs as, after further discussion and investigation it was considered that there might be a slight risk to personnel if the Halon gas escaped into the tunnel atmosphere. The TBMs were fitted with manually activated AFFF foam systems and were zoned by two water curtains, to control smoke movement within the TBM.

74. *Marine running tunnel TBMs* Fixed AFFF foam fire-fighting systems were installed on the machines, together with a water curtain, to assist evacuation in case of fire. Manual appliances were located throughout the TBMs. Halon was not used. A continuous environmental monitoring system was operated which also fed the information to the surface control room.

TBM erection and commissioning

75. All TBMs were factory erected and tested. They were then stripped down into component parts to allow them to be transported by way of the 10 m diameter access shaft and along twin 900 gauge construction railway tracks to the underground erection chambers. Before completion of the shaft, the marine service tunnel TBM components were introduced by way of the A1 Adit, connecting the 1974 section of the service tunnel to the Lower Site, and erected in an enlarged chamber created after the removal of the 1974 TBM cutterhead. The erection chambers for the running tunnel TBMs were specifically excavated for this purpose. They measured 13·4 m wide and 14·5 m high and were constructed using the NATM method (see Fig. 6).

76. The largest items transported into the erection chambers were the main bearings

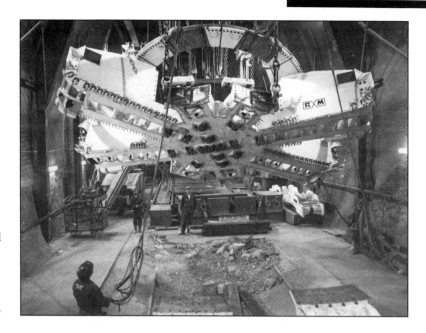

Fig. 7. Marine running tunnel TBM erection

which were up to 86 t in weight and of 6·8 m dia. Erection within the marine chambers was carried out using a series of 50 t hoists, mounted on a structure of steel arches and interconnecting monorail beams (see Fig. 7). However, it was found that there was insufficient flexibility with this arrangement and, therefore, the land TBMs were erected using overhead gantry cranes, supported by concrete sidewalls, the gantry cranes providing both longitudinal and lateral movement (see Fig. 8).

77. Owing to the long overall length of the running tunnel TBMs, only the machine and the conveyor section of the back-up could be constructed and commissioned before the launch. Even in their shortened form, the TBMs projected beyond the bunkers. Initial excavation, therefore, could not employ spoil trains. Spoil was transported on temporary extendable conveyors and was discharged directly into the marshalling area bunkers for the first kilometer of drive. After sufficient drive length had been established, installation of the remaining back-up wagons and equipment was completed and the whole machine fully commissioned.

78. The erection of the running tunnel TBMs took 14 weeks, using an average gang size of 30 men, working around the clock, seven days per week. As each TBM erection was completed, part of the team remained to start commissioning the TBM, and to form the nucleus of the mechanical and electrical gangs for that drive.

Tunnelling support services

79. The programme required a high-speed tunnelling operation, involving six TBMs driving long distances. This operation demanded a comprehensive and, inevitably,

complex system of support. The services to support the drives included spoil handling, water supply, drainage, electrical distribution, ventilation, communication, environmental monitoring and the creation of the largest construction railway in the world.

Construction pumping and drainage

80. The pumping system installed in the marine tunnels was designed to carry 645 l/s. To contain unexpected flows from faults or unsealed boreholes, the marine TBMs were equipped with operational and standby pumps capable of discharging to temporary pumping stations built in tunnel cross-passages, up to 3·5 km from each face (see Fig. 9).

Construction power

81. Six separate 11 kV feeder cables provided electrical power to the site. This was distributed from a primary substation to meet a maximum demand of 23 MVA. Diesel generators (four at 1·25 MVA) provided an emergency back-up for essential services. Separate 11 kV ring mains fed the TBMs and services in each marine tunnel (see Fig. 10). The land tunnels were supplied by radial feeders. Secondary substations in tunnel cross-passages provided 415 V services, at 1·125 km intervals along each tunnel. These provided power for lighting, pumping, ventilation, DC traction, signalling and communications.

Construction ventilation

82. Initially, ducted fresh air systems provided quantities in excess of 9 m³/min per m² of face area. Owing to the length of each tunnel, special consideration was given to duct friction losses and air leakage. The marine service tunnel, at 22 km, was the longest single line ducted ventilation system in the world.

83. The air was supplied through fire-resistant flexible ducting, of 2 m dia. in the running tunnels, and 1.2 m dia. in the service tunnels.

84. Two fan stations on the surface at Shakespeare Cliff Lower Site were supplemented in the tunnels by in-line fans, positioned at every ten cross-passages (3·7 km). These in-line fans were 90 kW variable speed units in the running tunnels and 60 kW fixed speed in the service tunnels.

85. At the surface fan stations, five banks of six fans, at 64 kW each, were employed at adit A1 portal, with one bank providing air to each running tunnel plus a standby. At adit A2, a smaller set-up provided air for the land and marine service tunnels.

86. Following breakthrough of the marine running tunnel TBMs into the crossover, the running tunnel ducted system was decommissioned and a 'mines type' system introduced (see Fig. 11). This technique used the tunnel bores as main arteries. Air was forced into the south tunnel by the adit A1 fan station at

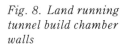

Fig. 8. Land running tunnel build chamber walls

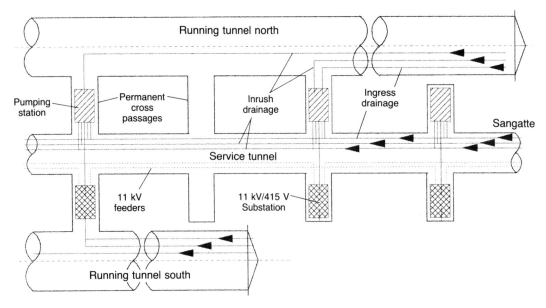

Fig. 9. Pumping arrangement

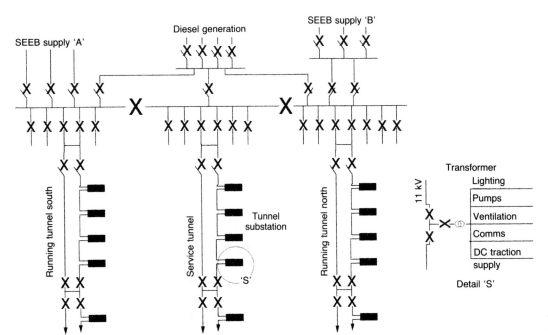

Fig. 10. 11 kV supply and distribution

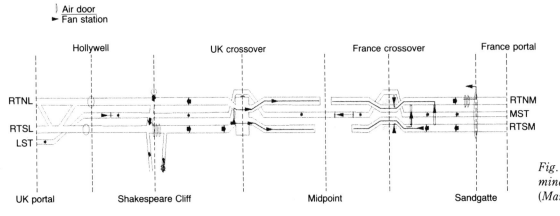

Fig. 11. UK–France mines type ventilation (March 1991)

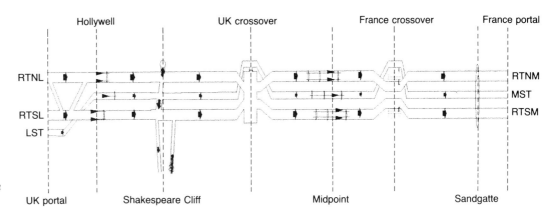

| Air door
► Fan station

Fig. 12. UK–France mines type ventilation (January 1992)

Shakespeare Cliff. It then travelled to the crossover, where it was then ducted to the TBMs before returning by way of the north tunnel and the service tunnel. Short circuit of the ventilation system was prevented by the installation of air locks in the south running tunnel at the Shakespeare Cliff marshalling area.

87. This system necessitated the sealing of all cross-passages and piston ducts in the south running tunnel between Shakespeare Cliff and the crossover. Several problems were presented by the need to seal the main spoil removal drift conveyor in adit A1, and also the various conveyor tunnels which honeycombed the Shakespeare Cliff marshalling area.

88. Remote control of the system at the Shakespeare Cliff main control room enabled maximum flexibility to deal with emergency situations.

89. After breakthrough of the marine tunnels to France, the system was replaced by fan stations at the Holywell shafts and at the marine tunnel mid-points (see Fig. 12).

90. This system of ventilation was designed as a unidirectional, high volume, pressure-balanced arrangement. The fans at the Holywell shafts forced fresh air into the complex at the UK portals, and, in conjunction with the mid-point fans, they induced an air flow through the tunnels which exhausted at the Sangatte shaft in France.

91. The Holywell fan stations employed five banks of two fans, at 60 kW each, in the running tunnels, and two banks of two fans, at 40 kW each, in the service tunnel. The mid-point fan stations employed four banks of two fans, at 90 kW each, plus two banks of three fans, at 40 kW each, in the running tunnels and service tunnel respectively.

92. Marine side air flows were typically 135 m³/s and 40 m³/s in the running tunnels and service tunnel respectively. These flows gave air velocities of up to 3·5 m/s.

Construction communications and monitoring

93. A seven-channel radio system formed the basis of the site communication system above and below ground. Back-up was provided by telephone, paging and public address systems. An auxiliary fire-safe telephone system was also provided.

94. The marine service tunnel system at 22 km had the longest multi-channel radio transmission system in the world.

95. All systems were self-monitoring and were connected to a Communications and Control Centre on the surface where the environment and all essental underground equipment was monitored on VDU screens. The Control Centre, which was continuously manned, also contained a major incident control room with facilities for the emergency services.

Environmental monitoring and telemetry

96. Evidence of a safe working environment in the tunnels, and at the work faces, was provided by a comprehensive arrangement of remote sensors, connected by means of a telemetry system to the Control Centre. These monitored the air temperature and the concentrations of oxygen, hydrogen, carbon monoxide, methane and oxides of nitrogen in the air. Alarms were set at levels which were designed to allow time for corrective action to be taken to prevent a situation developing where personnel and the tunnels would be put at risk

Logistics

97. *The railway* A basic principle of the 900 mm gauge railway was to use common rolling stock for all drives. This resulted in trains supplying the service tunnel with the segments and spoil capacity for two rings, and the marine running tunnels with the capacity for one ring. On the land running tunnels, a combination of limiting factors—including number and type of available locos, the larger

volume of excavated spoil per ring, the heavier rings and the geometry of the TBM conveyors and tracks—led to the decision to service these faces with trains having a capacity of only $\frac{1}{2}$ ring of excavated spoil.

98. The spoil cars were Mulhauser 14 m³ side tipping wagons. For the marine running tunnels, spillage boards were added to the wagons which increased their capacity to 15 m³. Although this was of advantage to the marine running tunnel drives, it compromised the principle of common rolling stock, as the 15 m³ skips were too high for use on the land running tunnel TBMs.

99. The intention was to use locomotives powered from 550 V overhead shrouded conductors for all drives. After overcoming considerable development problems with locomotive pantographs, ingress of saline water and driver training, this system was successfully used for the majority of the marine drives.[5] However, development of the ventilation systems allowed these to be supplemented with diesel locomotives, and the land running tunnels were serviced entirely by diesel locomotives.

100. *Bunkers and conveyors* The spoil removal system comprised a network of bunkers and conveyors, approximately 3 km in total length, designed to operate at a nominal capacity of 2400 t/h over a four-year period.

101. Each tunnel drive had its own independent moving bed bunker, mounted below rail level in the tunnel invert, located in the marshalling area at Shakespeare Cliff, and receiving spoil from the side-discharge muck skips of the spoil trains. The primary function of the bunkers was to receive up to 270 t of spoil, tipped in a period of approximately 30 s, and then to discharge it at a controlled rate of 400–600 t/h to the downstream conveyor system.

102. The underground conveyor system was contained within a network of interconnecting tunnels and galleries, constructed beneath the Shakespeare Cliff marshalling area (see Fig. 13).

103. The bunkers were based on a standard 110 kW National Coal Board design, employing a 30 mm round-link chain in a double moving bed arrangement. The drive to the chains was provided by a 110 kW hydraulic power pack by way of radial piston motors and helical reduction gear boxes. The bunker output was controlled by probes in the discharge chutes which stopped and started the hydraulic drives.

104. Each bunker discharged on to a metering chain conveyor, which regulated the flow of spoil before discharging, in turn, on to a downstream belt conveyor.

105. At peak tunnelling production, five bunkers were in operation simultaneously. The spoil from all five bunkers converged underground and was then brought to the surface by way of a 675 kW drift conveyor in adit A1. This main conveyor was rated 2400 t/h, with a belt speed of 3·7 m/s, and carried 117 t when fully loaded.

106. The adit A1 conveyor discharged on to a surface conveyor system totalling 1000 m in length, consisting of a series of six overland conveyors and a 70 m radial spreader conveyor, which discharged behind the seawall.

107. The reliable operation of the conveyor system was fundamental to the progress of the entire tunnelling operation. While a TBM malfunction stopped one tunnel drive, a conveyor malfunction could stop five tunnel drives. This applied equally to planned shutdowns of the system. Co-ordinated simultaneous stoppages of all TBMs were impractical to organize, which meant a convenient time to stop the conveyors was never available. Eventually, a conveyor maintenance hour was created and took place during the day/back shift-change period, to minimize impact on production.

108. Accordingly, the design of all conveyors included multiple or duplicate drives to enable production to be maintained (albeit at a reduced rate) in the event of a drive failure.

109. The entire system of 29 conveyors was managed by means of programmable logic controls (PLCs) and mimic boards, located in an office in the marshalling area. Remote TV cameras at each bunker assisted the conveyor controller in the selection of conveyor routes, appropriate to the tunnel drives in production at that point in time. The conveyors and bunkers could all be remotely stopped and started from the control office in the marshalling area.

110. With bunkers rated at 600 t/h and the main conveyor in adit A1 rated at 2400 + t/h, it was not theoretically possible to discharge all five bunkers simultaneously. Priorities for the order of bunker discharge, generally in favour of the marine drives, had therefore to be established. However, pressure of production often resulted in conveyor outputs well in excess of 3000 t/h.

111. The spoil itself proved to be a most difficult material for the conveyor system to handle. It varied enormously in density, moisture content, lump size and consistency.

112. The marine drives produced spoil which was generally well graded, with a high proportion of lumps under 100 mm. However, the moisture content varied, giving a consistency which ranged from dry and dusty to a slurry.

113. The land drives were generally more consistent, but generated spoil which, although apparently dry, was actually of a moisture content which produced the worst clogging effect in the conveyors. Towards the end of the land tunnel drives, the spoil became blocky

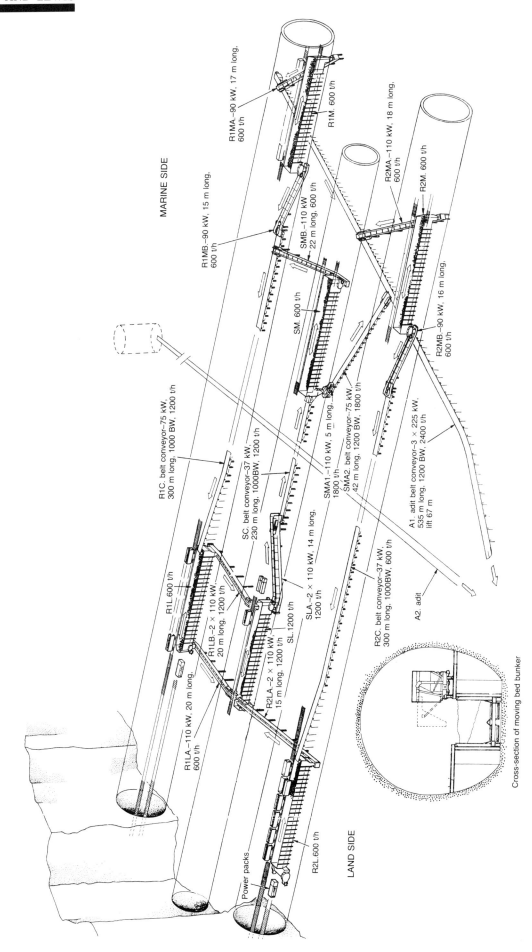

MARINE SIDE

R1MA.–90 kW, 17 m long, 600 t/h

R1M. 600 t/h

R2MA.–110 kW, 18 m long, 600 t/h

R2M. 600 t/h

R1MB.–90 kW, 15 m long, 600 t/h

SMB.–110 kW 22 m long, 600 t/h

SM. 600 t/h

R2MB–90 kW, 16 m long, 600 t/h

R1C. belt conveyor–75 kW, 300 m long, 1000 BW, 1200 t/h

SC. belt conveyor–37 kW, 230 m long, 1000BW, 1200 t/h

SMA1.–110 kW, 5 m long, 1800 t/h

SMA2. belt conveyor–75 kW, 42 m long, 1200 BW, 1800 t/h

A1. adit belt conveyor–3 × 225 kW, 535 m long, 1200 BW, 2400 t/h lift 67 m

R1L.600 t/h

R1LB–2 × 110 kW, 20 m long, 1200 t/h

SLA.–2 × 110 kW, 14 m long, 1200 t/h

SL.1200 t/h

R2LA–2 × 110 kW, 15 m long, 1200 t/h

R2C. belt conveyor–37 kW, 300 m long, 1000BW, 600 t/h

A2. adit

R1LA–110 kW, 20 m long, 600 t/h

Power packs

R2L.600 t/h

LAND SIDE

Cross-section of moving bed bunker

Fig. 13. Channel Tunnel underground conveyor system

and, as such, created problems, particularly for belt conveyors. The blocks—up to 1 m³ in size—if not broken up manually at the TBM, caused extensive damage to idlers and chutes and had to be broken up manually in the bunkers.

114. The spoil, in its various forms, produced a variety of problems to the belt and chain conveyors alike. The friction of the spoil against steel was higher than had been anticipated. This, combined with spoil build-up around chains, resulted in very high chain tensions and power requirements. This sometimes resulted in full bunkers unable to move, or worse still, full bunkers with broken chains. In such cases, the bunkers had to be dug out by hand before the chains could be repaired or replaced, and this could lead to considerable delays.

115. Strict procedures were therefore introduced to control muck skip discharge in order to maximize throughput, but to minimize downtime. These were often different for each tunnel and were varied to suit prevailing spoil conditions.

116. Over a period of time, the specifications for the chains and gear boxes were uprated. Ultimately, the configuration of bunker chains and flights was enhanced by 100%, from a double to a quadruple chain system, and incorporated an automatic pressure limiting feature.

117. The problems experienced with belt conveyors were as diverse as with the chain conveyors. Spillage from belts, either in dust form or droppings from the return belt strand, necessitated the employment of a large labour force to maintain cleanliness. Any residual spoil which remained on the belts would cause tracking problems and, in extreme cases, belt inversion. Belt scrapers, therefore, required careful adjustment to achieve the efficient removal of spoil without causing damage to belts and belt fastenings.

118. Another factor that affected the conveyor system, and that had a potential to cause major delay, was the weather. Heavy rain and high seas could render the surface conveyor system inoperable, despite all attempts to protect the conveyors. A standby fleet of dumper trucks, which could be used to transport the spoil in the event of surface conveyor breakdown, was equally vulnerable to weather. In the worst gale conditions, delays to tunnelling were inevitable.

Alignment

119. With such long drives, survey activities were crucial to successful completion. Overall management and quality control were managed by a combined UK/France Task Force, reporting directly to the Project Executive Committee.[9]

120. A fully triangulated network of stations was taken along both sides of each tunnel.[1] Space was extremely limited in the service tunnel and stations were so close to the tunnel walls that refraction effects were significant.

121. The use of gyrotheodolites with an accuracy of ±3 seconds of arc were vital in the successful tunnel breakthroughs subsequently achieved.

122. At the face, section engineers worked from the most forward main survey station to align lasers on to the ZED 260 systems built into each TBM. The ZED 260 systems were linked to personal computers (PCs) on the surface, where readouts could be obtained and survey information downloaded to computers in the TBMs.

Tunnel drives

Marine service tunnel

123. The drive began on 29 November, 1987, from the position where the 1974 Tunnel had been abandoned.

124. The marine service tunnel was driven almost entirely within the Chalk Marl zone, although occasionally the underlying greenish Glauconitic Marl rose into the bore. Some stretches of the drive enjoyed dry and relatively stable face conditions. However, for considerable lengths of the drive, blocky and often wet conditions prevailed.

125. The wet broken strata were first encountered in March 1988, with only 800 m of tunnel driven. These difficult ground conditions continued until December 1988, by which time only 3·5 km of the drive had been completed.

126. To deal with the difficult conditions, modifications were carried out on both the TBMs and many of their support systems. Of particular note was the addition of a flexible hood, made up of trailing fingers and thin stainless steel infill plates, spanning from the tailskin to the last ring. This was designed to allow safe erection of the unbolted ring. The hood contained the broken ground, reduced the overbreak and prevented unacceptable distortion of the expanded lining.

127. When the worst ground conditions had been overcome, the driving rates increased. In the latter part of the drive, weekly production figures generally exceeded 200 m and peaked at 293 m.

128. Despite several early attempts to use the bulk silos and transporters of the Schwing grouting system, practical considerations on the TBM, together with the damp atmosphere in the tunnel, led to a return to more traditional grouting methods, using pan-type paddle mixers and Craelius pumps.

129. A three-stage grouting system was

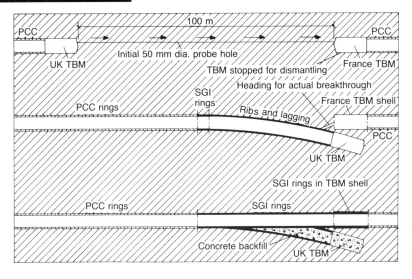

Fig. 14. Plan view marine service tunnel junction

evolved in order to overcome the physical restraints of access to grout holes when the TBM was advancing at rates of up to 3 m/h. In practice, grouting up to axis level was carried out within 8 rings of the build area. The tunnel above this was grouted from a platform, mounted above the conveyor and extending back a further 32 rings.

130. In the drier zones, bagged Pozament GP3 was generally used, while in the wetter areas, ordinary Portland cement (OPC) with an anti-washout additive was used. OPC was also used for proof grouting.

131. At the rear of the TBM gantries, purpose-made sledges were added to facilitate proof grouting. These sledges also allowed access for repairing and cleaning-down segments, to produce a 'finished' tunnel as it left the back of the machine.

132. The probing system, consisting of two Craelius drilling rigs, was provided at the front of the TBM and was designed to allow two 75 m long holes to be drilled in one 8 h shift. The drilling rigs were folded away under the conveyor head assembly during mining operations. The conveyor was withdrawn from the TBM face cone to give access to the face for the probing operation. Strict isolation procedures were enforced before any entry into the head. Drilling rods were aluminium, to allow, if necessary, for safe abandonment ahead of the TBM face.

133. Considerable practical difficulties were initially found with drilling the two long forward probes, resulting in large delays to production. A series of modifications were made during the early months of driving, resulting in the acceptance of a single forward probe up to 240 m in length. This was generally drilled within a single 8 h maintenance shift. Because the verticality of such an extended hole was very difficult to predict, a method of ascertaining the level (to within ±1 m) by

micropalaeontological analysis of drill flushings was developed by the site geotechnical team.

134. Rapid, reliable and informative forward probing was then maintained throughout the drive. Downward probes were drilled to locate the Gault Clay horizon, and thereby to establish the actual geological datum.

135. Sideways probe holes, for exploration of the geological conditions ahead of the running tunnels, were drilled only in zones where the service tunnel geology indicated a need for further information. These holes were cored and permeability tests carried out. In general these were within the acceptable limits of 3 lugeon. However, a 700 m length of the marine running tunnels was considered to require ground treatment when permeabilities in the range 3–29 lugeon were found (see §§ 171–179).

136. The cross-passages which housed temporary substations and stage pumping stations, were constructed from the service tunnel to within 2 m of the running tunnels. The construction of these cross-passages advanced with the tunnel drive, lagging some 2–3 km behind the advancing face. The remainder of the cross-passages, electrical rooms and signalling rooms were similarly constructed, but further behind the TBM 'construction zone', thus minimizing disruption to the logistics supporting the drive.

137. The construction of these latter passages, which was not essential for the tunnel drive requirements, was used to maintain an even workload for the mining crews involved.

138. Despite plans to employ a mechanized system of excavation for these structures, the disruption and restrictions on the service tunnel face production, imposed by such equipment, led to all cross-passages being excavated by traditional hand mining methods.

139. Other major structures excavated from the marine service tunnel during the course of the drive included the UK undersea crossover,[6,7] together with major excavation and construction works for the two main pump stations.[6] All material supply and spoil removal to and from these major works was by way of the service tunnel.

140. On 30 October 1990, in accordance with the original programme and at a position over 22 km from the starting point, contact was first made with the French service tunnel TBM by a 50 mm dia. probe hole through the tunnel face. The UK was therefore joined to the Continent by direct link for the first time since the English Channel was formed.[1,9]

141. Following initial contact with the French TBM and confirmation of the survey, the UK TBM was driven off line to a position adjacent to the French TBM (see Fig. 14). The first full contact with the French tunnel was by

way of a short lateral heading, driven into the face of the French TBM. The breakthrough took place on 1 December 1990.

142. Following removal of the TBM back-up equipment, the tunnel was completed by hand excavation and was lined with SGI rings. The TBM was entombed in concrete outside the finished lining. By the time the SGI lining was completed, the French TBM had been removed, leaving in place its outer skin. The SGI rings were built within this skin, to connect with the last concrete ring placed by the French TBM.

Land service tunnel

145. The 5·8 m dia., 8 km long, land service tunnel drive began on 1 November 1988. The tunnel was driven uphill on a gradient of 1 : 91, with the first 5·5 km on a 6 km radius (see Fig. 15). The tunnel was driven largely through the 10–15 m thick bed of lower Chalk Marl, and apart from some fracture zones within the first 2 km of drive, most of the tunnel was driven in the anticipated ground conditions. During the last kilometre of drive, however, glauconitic marl entered the face to give the most difficult conditions encountered during the drive.

144. During the drive, the rings were grouted in two stages behind the TBM. The first stage used a paddle-pan mixed grout (Pozament and water) which was injected, by means of a Craelius pump, through grout holes in the invert segments, to just above haunch level.

145. Second-stage grouting was carried out from a grout platform situated above the conveyor, 8–10 rings behind the last ring built, and the grout was injected in both shoulders and through the void between adjacent keys.

146. The material used for the upper grouting was a bulk batched Pozament, using a continuous process by a Schwing mixer and pump arrangement.

147. Any leaks or areas missed in these two stages were dealt with at the rear of the TBM, from a proof grouting platform. Here, the infill to key voids was methodically proof drilled and grouted, segments cleaned down and repaired to provide a 'finished' tunnel at the back of the sledges. Drilling rigs for probing were mounted in the TBM, although ground conditions never warranted their use.

148. Over the last 600 m of the drive, the glauconitic marl proved very broken and blocky, and also contained the first significant water encountered in the drive. Even though the water was in relatively small quantities, when it mixed with the chalk marl, a particularly puggy material was formed which blocked up the cutterhead hoppers, substantially increasing cutting times. When compacted, the material hardened rapidly and proved difficult to remove, resulting in long periods of downtime. On a number of occasions, the TBM head became completely blocked, requiring people to enter the head and clear the blockage by hand excavation.

149. The Glauconitic Marl, being blocky and broken, also resulted in large overbreak during excavation. This, when exposed during the ring building cycle, required extensive and meticulous timber packing to maintain correct ring profile. Large quantities of grout were subsequently injected into these areas.

150. Finally, on approaching the shaft at Holywell, the cover reduced to between 5 m and 10 m as the side of Sugar Loaf Hill was approached. After a number of face and roof collapses, the ground around the shaft, including part of the hill, was stabilized with concrete and grout. The final length of tunnel was lined with heavy duty SGI rings.

151. The drive broke through on 9 November 1988 at Holywell Coombe after just over 12 months' driving, the first of six UK TBM breakthroughs.

Land running tunnels

152. The land running tunnel north drive began on 2 August 1989, and the land south drive began on 27 November 1989. Both tunnels were driven on uphill gradients of 1 in 91 parallel to the service tunnel.

153. The boreholes indicated that ground conditions for most of the drive would be dry and stable Chalk Marl. However, it was known from the experience of driving the land service tunnel that roof conditions during the drive could be unstable. Consequently, because of the large diameter of the running tunnels and the larger span of unsupported roof in the build

Fig. 15. Land service tunnel plus train

area, and the fact that the roof of the excavation would extend further into the Glauconitic Marl horizon and encounter it more frequently, trailing ground support fingers were added to the tail of the shield. These limited the effects of overbreak and provided a safe environment for erection of the rings.

154. The segments in the land running tunnel were 540 mm thick. The heaviest of these was the invert segment, weighing 8 t, with the standard segments weighing 6 t. These heavy segments placed enormous loads on the telescopic erectors and erector heads. Although the erectors were designed to cope with heavy loads, in-service experience led to further design modifications and strengthening of the slider arrangements. The erector arms and heads were exchanged on a regular basis, to allow servicing and refurbishment. To minimize the delay arising from component failure or breakdown, a spare erector arm and head were carried on each TBM.

155. Following erection, the invert segments were grouted first. The profile of the grout face sloped upwards, away from the build area, to prevent grout losses. The crown and key voids were grouted from the grouting platform, about 9 m behind the leading ring. Early stabilization of the invert and knee segments was necessary to support the heavy load transferred by way of the front platform, and early striking times of the key void shutter also demanded fast setting times.

156. For the land running tunnel north, the grouting system adopted was designed by Fosroc Ltd. The system provided a retarded grout comprising Pozament GP3 and Conbex 802, mixed in twin 0.5 m^3 vertical pan mixers. These fed a 1 m^3 vertical agitator pan which discharged into a ringmain system installed around the top grouting deck. The retarded grout was circulated by mono pumps, and the grout was taken off at different points for injection at invert, axis or crown level.

157. A liquid accelerator, Conbex 803, was added, by means of a static in-line mixer, at the point of injection. A computerized system of control adjusted the dosage of accelerator in relation to the volume of grout flow, the rate of injection being measured by flowmeter. The dosage of the accelerator could be varied as a percentage of the mix, and thus vary the setting times, as required.

158. During the course of both drives, the face encountered a number of faults and the trailing fingers proved to be of great benefit in providing support to the broken ground in the roof above the build area. In such areas, overbreak of up to 1.3 m was packed with timber and dry concrete, before building of the ring.

159. In view of the adverse ground conditions encountered by the land service tunnel drive as it approached Holywell reception shaft,

20 m long TBM reception chambers were constructed from running tunnel reception shafts, using the NATM method. In addition, steps were taken to exclude the water and improve the stability of the ground beyond the end of the chamber by drilling and grouting from each head wall. A protective hood was formed over the crown of each tunnel, consisting of 30 m long by 75 mm dia. holes, through which cementitious grout was injected.

160. A decision was also taken to treat the ground for the last 0.5 km of each of the land running tunnel drives. This was carried out by drilling and grouting from the service tunnel, in advance of the running tunnel drives.

161. Using controlled pressures of up to 18 bar (the maximum overburden pressure), a claquage (hydrofracture) grouting technique was adopted to penetrate the fine fissures. A 20% bentonite/cement grout was injected through three holes drilled radially at 1.5 m intervals from the service tunnel. The holes were drilled in the form of a fan-shaped array, to provide a protective umbrella above the crown of each running tunnel in order to reduce the amount of water entering through the Glauconitic Marl.

162. The ground treatment proved to be of considerable benefit. Face and roof stability was improved by the successful exclusion of water. The ground treatment prevented the blocking of the head with Chalk Marl pug, as experienced in the land service tunnel, and therefore relieved the running tunnel drives from the arduous task of digging out the blocked head, by hand.

163. However, the blocky nature of the ground still caused severe difficulties for the workforce when the expanded lining was being erected. The broken ground imposed heavy loads on the trailing fingers and the overbreak had to be hand-packed with bags of concrete and timber. During this period of blocky ground conditions, the cutterhead of the TBM was maintained against the face, to provide support and to prevent blocks from slipping out of the face.

164. The potentially dangerous conditions, caused by the blocky and unstable nature of the face, sometimes prevented engineering staff from entering through the cutterhead into the face, to measure the cut diameter and to check the wear and to adjust the gauge picks. Postponement of checks on pick condition could lead to variations in cut diameter and, consequently, give rise to problems with the build of the ring and the correct choice of key size. It became necessary, therefore, in order to prevent delay, to carry keys of all three sizes at the face or on the material train.

165. Even when conditions were judged such that safe entry to the face could be achieved, temporary support was installed to

prevent blocks slipping out of the face and on to the cutterhead. In all cases, entry to the face had to comply strictly with safety procedures, including isolation of the head and the conveyor.

166. Workmanship was of a high order. The design of the machine enabled the workforce to gain safe access to all parts of the constructed lining, and allowed finishing crews to proof grout the key voids, to clean and carry out any repairs, and generally to provide a 'finished' tunnel as it left the back of the machine (see Fig. 16).

167. Junctions at the cross-passages, already excavated from the service tunnel, were completed behind the TBM construction zone. Also construction of the piston relief ducts was carried out concurrent with running tunnel driving, behind the TBM construction zone.

168. During the launch and the first 2 km of the land north drive, a few modifications and changes to the plant and equipment were required, but once these had been undertaken, progress was limited only by the logistics of the ring supply and spoil removal. At the time that the land running tunnel drives were in operation, enormous logistical demands were being placed on the contract as a whole. The supply of all materials for the simultaneous operation of five tunnel faces, together with cross-passage, piston duct, pump station and the undersea crossover construction, all competed for the service of the railways, bunkers and conveyors.

169. The land running tunnel drives frequently achieved a full ring cycle of excavate, build and grout in a 20 min period, shifts of 25 m being the best attained. However because of the demand for resources on the more critical marine drives, the full potential of the land running tunnel TBMs was never realized.

170. The land running tunnel north drive breakthrough occurred on 11 September 1990, with the south following on 20 November 1990. The TBMs and back-up equipment were dismantled and removed via the Holywell shafts.

Marine running tunnels

171. The erection of the TBMs was completed and the first ring was built in the marine running tunnel north on 15 March 1989, and in the marine tunnel running south on 19 June 1989.

172. The precast segmental rings had an o.d. of 8·36 m with segment thicknesses of 270 mm and 360 mm for use in different ground loading conditions.

173. Heavy duty SGI linings were also available and were used for the first sections of each tunnel: at the entrance and exit to the crossover cavern and adjacent to the main pumping stations. They were also used in areas

Fig. 16. Land running tunnel north

of bad ground where the concrete rings were unsuitable.

174. For a considerable length of the tunnel drives there were problems arising from overbreak in the crown, in some areas creating voids which extended up to 3 m above the excavated profile. High water inflows also occurred, causing serious problems with the TBM electrical systems that resulted in substantial delays.

175. The lining design anticipated the injection of 20 mm of grout around each ring. However, in the areas of high overbreak, the quantity of grout injected was significantly increased. A grouting system was adopted similar to that described for the land running tunnel north (see §§155–157) but, as in the marine service tunnel, bulk handling equipment was unsuccessful, on account of the humid conditions. The grout material used was therefore bagged Pozament GP3, combined with Fosroc retarding and accelerating agents. The setting time of the grout could be varied down to about 20 s depending on the ground conditions. In areas of high water inflow, the system provided an effective primary grout seal. Additional proof grouting and sealing was carried out 150 m back along the gantries.

176. Experience in the marine service tunnel had indicated particularly unfavourable ground conditions between chainages 22 700 and 23 400, with very wet and blocky ground. As the drive approached this zone, the indications were that conditions would pose an unacceptable risk to the programme performance of the marine running tunnels. Stage packer tests carried out in side probe holes, drilled from the service tunnel to running tunnel crown, con-

*Fig. 17. Finishing
works*

firmed this, with indicated permeabilities of
8–19 lugeons.

177. *Ground treatment* To improve the
ground conditions ahead of the running tunnel
machines, grout injection through holes drilled
from the marine service tunnel was employed.[8]
The grouting was designed to give a 3 m
annulus of treated ground around the upper
half of both running tunnels, and had to be
carried out with minimum interference to the
service tunnel operations.

178. By using separate rail-mounted drilling
and grouting trains, subcontractor, Stent-
Soletanche JV, drilled injection holes 13–21 m
long, forming a fan-shaped array at 3 m intervals through the treatment zone. Grouting was
by the tube *à manchette* method, and employed
a low viscosity, silicate based grout injected
under a carefully controlled pressure, which
just induced hydrofracture. All the relevant
parameters were monitored and logged auto-
matically.

179. The volumes of grout injected, the need
for controlled hydrofracture and the use of tube
à manchette had been determined by full size
trials, carried out in a cross-passage some
months previously. Post-treatment packer tests
indicated that permeabilities had typically
reduced to less than 1 lugeon, and monitoring
of the running tunnels though the treated zones
recorded minimal water inflows. While predicted advances in the untreated ground had been

85 m per week, the average advance rates
actually achieved through this area were
210 m/week, indicating the success of the
ground treatment.

180. *TBM modifications* To cater for the
adverse ground conditions and high water
inflows, the marine running tunnel TBMs
required extensive modifications to enable them
to work effectively. The major modifications
were as follows

(*a*) removal of upper slat segment conveyors to
make room for an upper grouting platform
directly behind ring erection area

(*b*) removal of twin lower slat segment conveyors and replacement with extended pawl
magazine system, to reduce downtime

(*c*) removal of lower segment erectors and con-
version of upper erectors for full 360°
movement

(*d*) major modifications to electrical units and
panels to protect them from water ingress

(*e*) removal of one probe rig and partial demobilization of the other, although this was
left in the head for possible use in emergencies, following the decision that sideways probes from the service tunnel would
be carried out in lieu of face probes from
the running tunnel

(*f*) major modifications to the computer
control systems linking the various TBM
and back-up systems.

181. *Follow-up activities* Once the very worst of the adverse ground conditions had been passed and the TBM modifications completed, segment cleaning and repairs were carried out to produce a 'finished' tunnel, at the back of the machine, for the remainder of the drives.

182. Seepage control works were carried out using purpose-built gantries. The gantries ran on the haunch nibs at a sufficient distance behind the TBMs to ensure that initial settlement had taken place and that any potential self-sealing had occurred (see Fig. 17).

183. Seepage control measures were designed to divert any ingress water by way of the circumferential joints into the permanent drainage system contained within the trackbed concrete. While the tunnels were not watertight, no free flowing water was permitted and no drips were allowed adjacent to areas of sensitive equipment, e.g. overhead catenary equipment.

184. As soon as possible after a running tunnel TBM had passed a cross-passage housing a service tunnel substation or stage pumping sump, the junction with the cross-passage was constructed and the passages completed. This enabled the running tunnel TBM services to be linked into these installations.

185. Remaining cross-passages, and electrical and signalling rooms were similarly completed, clear of the TBM construction zone. The pressure relief ducts (PRDs) were constructed by working alternately from each running tunnel. The setting out of the PRDs was complex and critical, and the valve was contained at the end. Therefore, excavation always began at that end.

186. *Passage through the crossover* The excavation of the crossover cavern was carried out from the marine service tunnel, ahead of the running tunnels.[6,7] The excavated cavern, which is 21·2 m wide, 15·4 m high and 164 m long, was completed down to cavern floor level by August 1990, when the north drive TBM broke through.

187. As the north TBM was the first to reach the crossover, its advance through the zone involved building a 117 m long section of the heavy duty bolted SGI lining, both approaching and leaving the cavern. In these areas, the tunnels had only 2–4 m of ground between them. It was therefore decided that when the south TBM passed these sections of tunnel, the SGI lining would be propped and anchored using steel and fibreglass rockbolts. Longitudinal fibreglass bolts were used to strengthen the 2 m wide rock pillar between the two running tunnels, adjacent to the cavern. Special measures were adopted to ensure the south TBM advanced rapidly and with reduced thrust and gripper pressures through these areas.

188. Deformations were monitored as the south TBM advanced past the reduced pillar width. The maximum deformation recorded was 10 mm, which was less than predicted. Grouting on the south running tunnels was carried out as soon as possible after ring erection in this zone. A short 21 m long section of heavy duty cast iron was used in the south tunnel at each end of the crossover.

189. The TBMs were shoved through the cavern, using the bottom auxiliary rams thrusting against temporary invert segments. Considerable maintenance work was carried out to the cutterheads and back-up gantries during the 7 day passage through the cavern (see Fig. 18).

190. *Progress* Once the TBMs and support

Fig. 18. Marine running tunnel TBM in crossover

equipment had passed beyond the influence of the continuing crossover construction, advance rates of over 340 m per week were achieved consistently by both TBMs, right up to the point of junctioning with the French. The north TBM advanced an average of 317 m per week between 17 September 1990 and April 1991, and the south TBM advanced an average of 321 m per week between 1 October 1990 and May 1991.

191. The drive lengths of the two tunnels were 17·36 km for the north tunnel and 18·77 km for the south tunnel. During the drives, record rates of progress were achieved, as follows: best week—428 m; best month—1718 m.

192. Each drive was completed with a short length of heavy duty SGI rings to facilitate the completion of the junction with the French tunnels. The last ring in the north drive (No. 11 716) was built on 23 April 1991, and the last ring in the south drive (No. 12 474) was built on 28 May 1991.

193. *Junction with the French* The method adopted for junctioning the British running tunnels with the French tunnels was to drive the British TBMs downwards on a 300 m radius vertical curve. Ground support through this section was carried out, using temporary segments in the invert, and shotcrete and fibre-glass rockbolts over the crown.

194. Once the TBM carcass was below the permanent tunnel line, the back-up gantries were removed and the tunnel was backfilled with approximately 2000 m³ of low strength concrete. This allowed the French TBM to drive over the abandoned British TBM, to a position within 1·0 m of the last permanent SGI ring (see Fig. 19).

195. The French TBM was then dismantled, leaving its skin intact. To connect with the last ring built by the French TBM, the short length of British SGI lining was extended through the

remaining skin of the French TBM. A short in situ section completed the junction.

TBM drive progress rates and records

196. *Average progress rates* All six TBMs met the required specifications, albeit after some modification of the marine machines to make them suitable for the difficult ground conditions encountered in the first 5 km.

197. The differences between the average progress rates for each drive (see Table 4) can be understood when the effect of all the variables encountered by each drive are taken into account.

198. The land tunnels (8 km) were shorter than the marine drives (18 km) and, consequently, their average rates of progress were more sensitive to delays. Therefore, the impact of delays caused by the learning curve and adverse geological conditions had a more significant effect on the land drive averages than on the marine drive averages. Although the marine drives encountered more severe ground conditions than the land drives, and the marine TBMs underwent modification, the considerable length of the marine drives diluted the effect of the delays.

199. There were, however, other particular factors, unique to the Channel Tunnel drives, which ultimately had a more significant effect on the progress rates for each drive and can be better understood when considered in the context of the overall project strategy.

200. The marine drives were known to be critical to the programme and, therefore, were given the highest priority in the allocation of resources, although a careful balance had to be achieved to maintain progress on the other drives.

201. It was not easy to achieve this balance, particularly from late 1989 to mid 1990, when five tunnel drives were in keen competition for the services of the railway organization. This occurred before plant resources had reached their maximum levels and, consequently, it placed enormous demand on the management of railway organization and severely complicated the logistics of providing an efficient service to all five faces.[5] A still-developing railway, bunker and conveyor system managed to cope, but not without some arguments and compromises.

202. At a time, therefore, when the land drives were beyond their learning curve and about to achieve their maximum rates of progress, the competition for resources was at its highest and, although they were able to meet and exceed programme rates, they were never allowed to realize their full potential.

203. On completion of the land drives, and following the marine service tunnel break-through, all the bunkers and conveyors, and a

Fig. 19. Section view of marine running tunnel junction

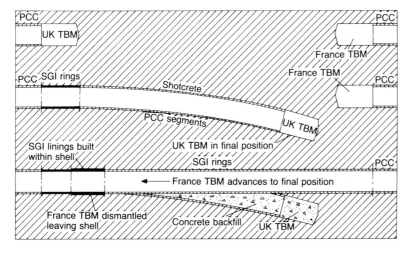

Table 4. Bored tunnel progress

	1986	1987	1988	1989	1990	1991
Marine service tunnel		29 Nov.		142 m/week	30 Oct.	
Land service tunnel			30 Sept.	137 m/week 09 Nov.		
Marine running tunnel north				27 Feb.	156 m/week	23 Apr.
Marine running tunnel south				30 May	179 m/week	28 May
Land running tunnel north				02 Aug.	134 m/week 11 Sept.	
Land running tunnel south				27 Nov.	151 m/week 20 Nov.	

large proportion of the railway resources, became available for the support of the marine drives. Consequently, just when the marine drives were extending their lines of supply and communication, beyond their economic/ logistical limit, a second set of resources was injected into the system.

204. A fully developed twin-track railway, bunker/conveyor system and pit-head organization, were able to provide unprecedented levels of service to the two marine drives and assist them in achieving their remarkable rates of progress.

205. A further conclusion may also be drawn if the individual averages for progress achieved by the north and south land drives are compared (and likewise a comparison made between the north and south marine drives). It can be seen, from examination of the records, that the north drives (which started first, carried the heaviest burden of learning curve, and suffered early on from shared resources) did not achieve the higher rates of progress of their almost 'identical' twin south drives. This is because the south drives started and finished last, and therefore benefited from the shorter

Table 5. Progress records

Description	MST	LST	MRTN	MRTS	LRTN	LRTS	Totals
Best shift (8 hours)	24 m 4 Feb. 1989	21 m 30 Mar. 1989	30 m 1 Feb. 1991	30 m 22 Feb. 1991	21 m 30 Mar. 1990	25·5 m 7 June 1990	
Best 24 hours	60·1 8 Feb. 1989	56 m 17 Apr. 1989	70·9 m 31 Jan. 1991	75·5 m 24 Feb. 1991	52 m 23 Apr. 1990	59·4 m 8 June 1990	214·3 m 23 Apr. 1990
Best week—Monday to Sunday	293 m 5 Feb. 1989	267 m 14 May 1989	409 m 10 Mar. 1991	426 m 24 Mar. 1991	308 m 15 Apr. 1990	320 m 10 June 1990	1210 m 6 May 1990
Best calendar month	997 m Apr. 1989	923 m Apr. 1989	1637 m Mar. 1991	1718 m Mar. 1991	1043 m Mar. 1990	1222 m June 1990	4481 m Mar. 1990
Best project month	1042 m Apr. 1989	1007 m Jul. 1989	1862 m Mar. 1991	1911 m Mar. 1991	1178 m Apr. 1991	1164 m June 1990	5170 m Apr. 1990
Achieved average m/week in in UK: m	142	137	156	179	134	151	
Achieved average m/week in France: m	111	72	149	159	77	68	

learning curves and, ultimately, from additional resources at the later stages of their tunnel driving.

206. *Progress records* During the progress of every one of the tunnel drives, world record advance rates were claimed which were then subsequently surpassed by the same drive or other drives on a number of occasions.

207. A schedule of the best advance rates of each of the drives are included in Table 5. All other final drive data are shown in Table 6.

208. *Planned against actual progress* The French tunnel drives also exceeded their anticipated progress rates by a considerable margin. This, in conjunction with the record British progress, meant that all tunnelling work was successfully completed on or ahead of the original 1988 programme, as shown in Fig. 20.

Channel Tunnel picks

209. The picks were generally of high tensile carbon manganese forged steel with tungsten carbide inserts, and were produced in varying grades of toughness and hardness. The rake angle of the tungsten carbide tip, together with the angle of pick 'rock', was designed to provide the most effective cut of the Chalk Marl.

210. Pick repair, by brazing new inserts into the machined face, was not pursued for technical, economic and progress reasons, and total pickhead replacement was adopted.

211. The direction of cut on the service and land running tunnel TBMs was regularly reversed to maximize wear time and to lengthen service life. However, this was not available to the marine running tunnel TBMs owing to the rocking pick configuration.

Table 6. Summary of drives data

Description	MST	LST	MRTN	MRTS	LRTN	LRTS	Totals
Cut rock date	29 Nov. 1987	30 Aug. 1988	27 Feb. 1989	30 May 1989	2 Aug. 1989	20 Nov. 1989	
Start of driving date (first cut)	4 Jan. 1988	30 Aug. 1988	27 Feb. 1989	16 June 1989	2 Aug. 1989	27 Nov. 1989	
5 km from adit A2	27 Jan. 1989	N/A	12 Apr. 1990	N/A	N/A		
End date (last ring built by machine)	30 Oct. 1990	9 Nov. 1989	22 Apr. 1991	17 May 1991	11 Sept. 1990	20 Nov. 1990	
Trailing edge ring 1: ch	19 841	19 075	19 571	19 531	19 072	19 004	
Leading edge last ring: ch	41 611	11 145	37 222	38 334	11 161	11 145	
Driven distance: m	21 770	7930	17 651	18 803	7911	7859	81 924
Rings built	14 478	5252	11 716	12 474	5240	5207	54 367
Distance from adit A2: m	22 310	8156	17 921	19 033	8140	8156	
Length of crossover	N/A	N/A	162 m/108 R	162 m/108 R	N/A	N/A	
PCC rings (m/rings)	13 363	5027	10 712	11 670	4925	4918	50 615
Opening set rings	521	163	319	321	148	148	1620
PRD opening set rings	N/A	N/A	224	234	92	92	642
Heavy duty CI rings	394	62	354	141	75	49	1075
Opening set numbers	115	36	77	80	36	36	380
PRD opening set numbers	N/A	N/A	71	76	31	31	209
Cut diameter: m	5·38	5·76	8·36	8·36	8·72	8·72	
Internal diameter: m	4·8	4·9	7·6/7·78	7·6/7·78	7·6	7·6	
Segment thickness: mm	270	410	270/360	270/360	540	540	
Ring length: m	1·5	1·5	1·5	1·5	1·5	1·5	
Volume of rock/ring: m³	34·10	39·09	82·33	82·33	89·58	89·58	
Total volume of rock: m³	494 895	206 637	983 704	1 044 743	472 448	469 343	3 672 770
Bulking factor on excavation	1·9	1·9	1·9	1·9	1·9	1·9	
Bulking factor as compacted behind seawall	1·24	1·24	1·24	1·24	1·24	1·24	

212. Pick consumption varied significantly in the changing ground conditions, but was generally low overall, with figures of approximately 500 m³ per pick (one pick per ten rings of tunnel excavated).

TBM performance monitoring

213. It was recognized at an early stage that it would be necessary to have good information regarding the TBM's performance on which management could base decisions. Traditionally, data for TBM drives are recorded in the form of shift reports, which tend to be either lacking in vital data or so voluminous as to be unusable. Babtie Shaw & Morton were commissioned to develop, install and maintain a suitable computer-based tunnel reporting/project management system, working closely with the tunnelling management on site.

214. The system was designed to overcome the traditional shortcomings of TBM reporting by collecting a large number of data in a simple format and then processing them by means of a bespoke computer programme. This was then used to output simple graphical reports that could be easily assimilated by busy managers. The reporting was achieved by the use of codes for all activities and possible down-time causes, facilitated by the use of a code matrix called Black Ball Charts (see Fig. 21). This enabled over 150 valid codes to be available from a simple alpha numeric grid.

215. The shift activity report consisted only of the activity start time and a single alpha numeric code. Input to the computer was then made simple by the use of entry screens that replicated the shift report format. Complex graphic reports were then generated by the computer using a report menu. The Artemis written computer programme was installed on the contract mainframe computer, the basis of the programme being a database, linked to the Artemis graphics. The system was extremely user friendly, and consistently produced the necessary reports with the minimum of input.

216. The reports generated were all graphical and produced on an HP colour plotter, A4 size. The reports included weekly and monthly progress statements, trend analyses, and the facility to report on specific areas. Detailed weekly reports were available within hours of the end of the week (see Fig. 22).

217. Despite the facility of the system to handle data and the sophisticated methods of data presentation, the process was still vulnerable to the 'rubbish in, rubbish out' syndrome. Education and training was, therefore, given to the engineering staff to ensure that they understood the value of the system, and vigilance was required to maintain a 'standard' approach to the recording and coding of events.

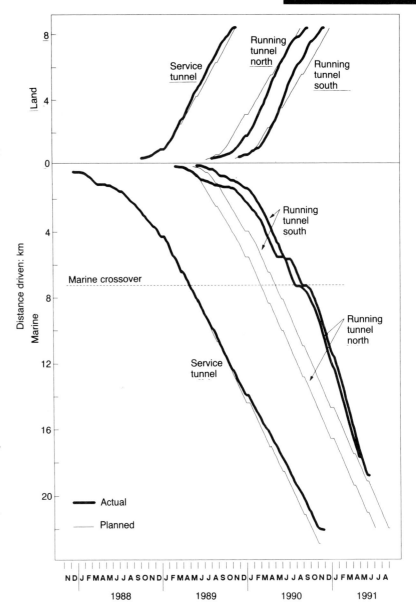

Fig. 20. UK tunnel drives planned vs actual progress

Maintenance management

218. To maximize TBM availability, it was recognized that maintenance (routine, special and emergency) were all required to ensure efficient operation of the TBMs.

219. During the time that five TBMs were driving concurrently, and with a working week made up of 24 hours per day, seven days per week, each TBM was shut down for an eight-hour maintenance shift once per week, on a rota system Monday to Friday dayshift. A special 'hit squad' as available on a regular dayshift to augment the squads dedicated to each particular TBM.

220. The maintenance shutdown periods also gave the opportunity for any special construction along the tunnel or placement of rail crossings etc. without disrupting progress of the drives.

221. In the case of the land drive TBMs, maintenance shifts were also arranged to coin-

Fig. 21. Black ball charts

Group	Code	Main category \ Sub-category	Normal operation 1	Electrical breakdown 2	Mechanical breakdown 3	Hydraulic breakdown 4	Lubrication failure 5	Bad ground 6	Supply error 7	Alignment 8	Miscellaneous sub-categories
Pro-duc-tion	A	Cutting		●	●	●	●		●		
	B	Re-grip machine		●	●	●	●				
Planned delays	Ca	Building rings outside cutting: Concrete	●								(Only Code 'C', if not, any other category below) 9 — bad build
	Cb	Building rings outside cutting: Opening	●								
	Cc	Building rings outside cutting: HDCI	●								
	D	Probing								●	9 — Set-up 10 — Grouting probes
	E	Planned maintenance		●	●	●	●	●	●	●	
	Fa	Cutter changes		●	●	●	●	●	●	●	
	Fb	Face inspection		●	●	●	●	●	●	●	
	Fc	Cable extension		●	●	●	●	●	●	●	10 — Ventilation cassette
TBM downtime	Ha	TBM breakdown Head rotation	●						●	●	10 — Human error (Ha to M)
	Hb	TBM breakdown Thrust/Grip	●						●	●	
	Hc	TBM breakdown Fingers/Shield	●						●	●	
	Hd	Cooling water system	●						●	●	
	Ia	TBM conveyors (fixed)	●						●	●	9 — Belt
	Ib	TBM shuttle conveyors	●						●	●	
	Ja	Lower segment cranes	●					●	●	●	
	Jb	Upper segment cranes	●					●	●	●	
	Jc	Upper erectors	●						●	●	
	Jd	Lower erector	●						●	●	
	Je	Upper slat conveyor	●					●	●	●	9 — Chain
	Jf	Lower slat conveyor	●					●	●	●	
	K	ZED/Survey equipment	●		●	●	●		●		
	L	TBM miscellaneous	●								
	M	Material unloading crane				●	●	●		●	
Other delays	N	Pump failure/Flooding	●			●	●		●	●	
	O	Engineer/Survey		●	●	●	●				
	P	Clean build area		●	●	●	●		●		
	Qa	Waiting for skips		●	●	●	●	●		●	9 — Derailment
	Qb	Waiting for segments		●	●	●	●	●		●	
	Qc	Waiting for other material		●	●	●	●	●		●	
	R	Cavity grouting									
	S	Service failure	●					●	●	●	
	T	Crew absent		●	●	●	●	●	●	●	
	Ua	Plant installation and commissioning									
	Ub	Plant modification		●	●	●	●	●		●	
	V	Miscellaneous other									
	W	Site closed		●	●	●	●	●	●	●	
	S	Missing record		●	●	●	●	●	●	●	

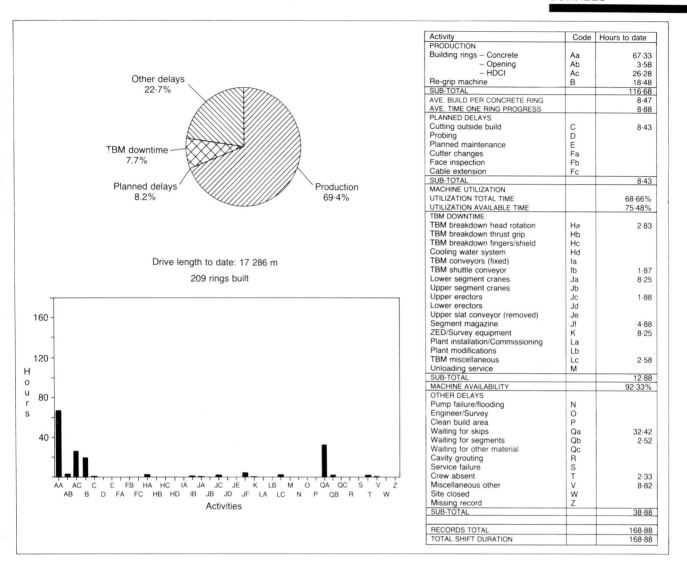

Activity	Code	Hours to date
PRODUCTION		
Building rings – Concrete	Aa	67·33
– Opening	Ab	3·58
– HDCI	Ac	26·28
Re-grip machine	B	18·48
SUB-TOTAL		116·68
AVE. BUILD PER CONCRETE RING		8·47
AVE. TIME ONE RING PROGRESS		8·88
PLANNED DELAYS		
Cutting outside build	C	8·43
Probing	D	
Planned maintenance	E	
Cutter changes	Fa	
Face inspection	Fb	
Cable extension	Fc	
SUB-TOTAL		8·43
MACHINE UTILIZATION		
UTILIZATION TOTAL TIME		68·66%
UTILIZATION AVAILABLE TIME		75·48%
TBM DOWNTIME		
TBM breakdown head rotation	Ha	2·83
TBM breakdown thrust grip	Hb	
TBM breakdown fingers/shield	Hc	
Cooling water system	Hd	
TBM conveyors (fixed)	Ia	
TBM shuttle conveyor	Ib	1·87
Lower segment cranes	Ja	8·25
Upper segment cranes	Jb	
Upper erectors	Jc	1·88
Lower erectors	Jd	
Upper slat conveyor (removed)	Je	
Segment magazine	Jf	4·88
ZED/Survey equipment	K	8·25
Plant installation/Commissioning	La	
Plant modifications	Lb	
TBM miscellaneous	Lc	2·58
Unloading service	M	
SUB-TOTAL		12·88
MACHINE AVAILABILITY		92·33%
OTHER DELAYS		
Pump failure/flooding	N	
Engineer/Survey	O	
Clean build area	P	
Waiting for skips	Qa	32·42
Waiting for segments	Qb	2·52
Waiting for other material	Qc	
Cavity grouting	R	
Service failure	S	
Crew absent	T	2·33
Miscellaneous other	V	8·82
Site closed	W	
Missing record	Z	
SUB-TOTAL		38·88
RECORDS TOTAL		168·88
TOTAL SHIFT DURATION		168·88

Other delays
22·7%

TBM downtime
7·7%

Planned delays
8·2%

Production
69·4%

Drive length to date: 17 286 m

209 rings built

*Fig. 22. Typical
weekly tunnel report*

cide with stoppages for 11 kV cable changes, and additional diesel generators were installed on the TBMs to enable work to be carried out independent of the tunnel power supply.

222. Routine maintenance, while the TBM was driving and during the eight-hour shift shutdown, was undertaken by the teams dedicated to each of the TBMs. The scope of the routine maintenance was determined by the suppliers. Job cards for performance of the work were issued weekly, with returns added daily to the 'Comac' planned maintenance computer system. This information also provided a database for major or special maintenance.

223. Maintenance of the bunker and conveyor route, dedicated to a particular tunnel, was carried out by conveyor crews on the same shift as the TBM was undergoing maintenance. Sections of the conveyor system which were common to all tunnel drives could not be closed down for maintenance without the stopping of all production. Maintenance of these sections, therefore, had to be carried out 'on the run' or in an agreed one-hour period at the day shift/

back shift changeover.

224. Special and emergency maintenance for TBM or conveyors was undertaken by a combination of the routine maintenance teams and the 'hit squad'. The nature of this work was usually major replacement of equipment or major planned modifications.

225. Special crews also visited each TBM both to monitor the maintenance routine checking and sampling oil, the lubricant levels, the fire protection, etc., and to report on trends and potential problems, indicated by analysis of the samples.

Gang size and labour costs for TBM drive tunnels

226. In order to maintain production, 24 hours per day, seven days per week, it was necessary to have four crews for each TBM. Generally a six on/two off shift pattern was worked, with three crews working and one off. Crews worked eight hours per day, changing strictly at the face. This arrangement was maintained even when the drives were at sig-

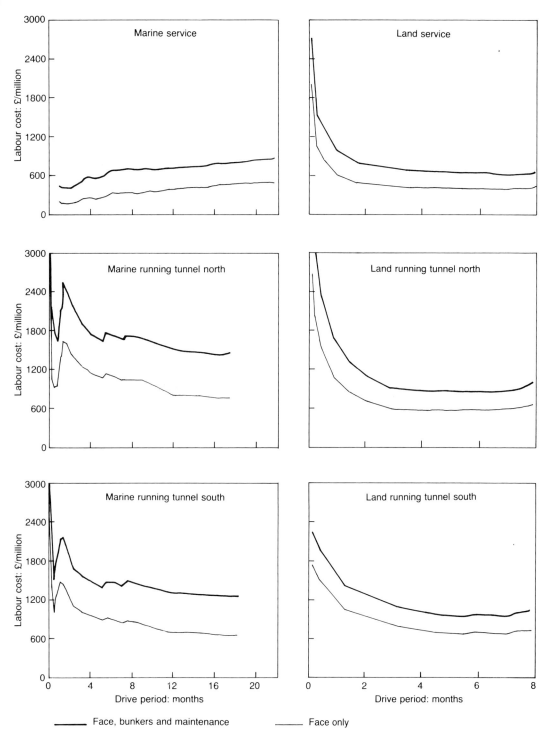

*Fig. 23. Comparison
of driving period and
unit costs/m of tunnel*

nificant length, when a 'travelling time'
payment was made.

227.   Crew sizes varied in accordance with
the particular requirements of each TBM (see
Table 7).

228.   Labour costs for the TBM drives were
recorded and monitored in several different
cost centres and categories.

(*a*) *Direct costs:* face crews; maintenance of

face equipment; marshalling area and
bunker crews

(*b*) *Logistical support:* underground operations
including rail transport; pithead; plant
maintenance; linear support services

(*c*) *Infrastructure:* Site service labour

229.   The graphs shown in Fig. 23 give an
indication of labour cost per metre for each of
the six TBM drives, identifying separately the

total direct cost and also the cost for the face crews only.

230. The levels of cost shown were current at the time of recording and so include escalation during the period of tunnelling.

231. Variations in cost per metre during the period of construction can be attributed to the following

(*a*) build-up to optimum numbers of operatives
(*b*) output achieved (ground conditions, machine performance, priority of resource allocation, learning curve, etc.)
(*c*) application of incentive schemes
(*d*) escalation
(*e*) logistics and travelling time.

232. Inspection of the cost records also shows that the ratio of logistical support plus infrastructure labour to direct labour cost was in the order of 2 : 1. Therefore, total labour cost is approximately three times direct cost.

Conclusions

233. Following a difficult start, caused by extremely adverse ground conditions, and after ground treatment and modifications to the marine TBMs to combat the conditions, the marine drives had fallen six months behind programme by September 1990. By sheer hard work and determination, the project recovered to achieve average progress rates of 320 m/week for 8 months for both marine drives and jointly with the French, to complete all tunnels on or ahead of the 1988 original programme.

234. Motivation of the workforce was of a very high order, there was very little turnover of labour, and a combination of commitment and enthusiasm permeated the whole workforce, from the directors to the hygiene squad.

235. Team spirit was essential to the success of the project and was encouraged through friendly competition, not only between crews/shifts/tunnels but also between the UK and France.

236. Against a backdrop of high quality and productivity, a safety management system was developed and refined to be the best in the industry.

Acknowledgements

237. The Authors acknowledge the assistance of their colleagues in TML in the preparation of this Paper. They also acknowledge the contributions of A. R. Biggart, Operations Director, and C. Pollard, Engineering Manager, and the assistance of D. Fawcett, of Babtie Shaw and Morton, for his contribution on TBM Performance Monitoring.

238. Finally, the Authors congratulate all who participated in this historic project.

References

1. VARLEY P. *et al.* Alignment and survey. *Proc. Instn Civ. Engrs, Civ. Engng, Channel Tunnel: Part 1 — Tunnels*, 1992, 43–54.

Table 7. TBM manning levels

Title	MST	LST	LRTN	LRTS	MRTN	MRTS
TBM pit boss	1	1	1	1	1	1
TBM operator/electrician	1	1	1	1	1	1
Leading miner	1	1	1	1	1	1
Miners	4	2	4	4	5	5
Crane drivers	2	2	2	2	3	3
Banksmen/Sleepers/Rails	2	2	1	1	1	1
Shuttle operators	—	1	1	1	2	2
Chargehand grouter	1	1	1	1	1	1
Front grouters	4	2	6	5	10	10
Proof grouters	2	2	2	2	2	2
Segment repairers	—	1	2	2	4	4
General operator/segment cleaner	1	2	1	1	2	2
Chargehand fitter	1	1	1	1	1	1
Fitters (incl. RMJV)	2	2	2	3	3	3
Chargehand electrician	1	1	1	1	1	1
Electricians (incl. RMJV)	2	2	2	3	2	2
Services fitters' mates	4	1	—	—	2	2
Services elect mates	2	1	—	—	2	2
Shunter/Banksman/Marshall	1	1	1	1	1	1
General operator/Spillage	—	—	1	1	1	1
Stage II Spec Finisher	1	—	1	1	1	1
Total	33	27	32	33	47	47

2. BRIDGES D. G. *et al.* Criteria and design methods for alignment of the Channel Tunnel—UK Sector. *Proc. Int. Symp. Tunnelling '91*, London, 1991.

3. EVES R. C. W. and CURTIS D. J. Tunnel lining design and procurement. *Proc. Instn Civ. Engrs, Civ. Engng, Channel Tunnel: Part 1—Tunnels*, 1992, 127–143.

4. CURTIS D. J. *et al.* The Channel Tunnel: design, fabrication and erection of PCC linings. *Proc. Int. Symp. Tunnelling '91*, London, 1991.

5. POLLARD C. *et al.* Construction planning and logistics. *Proc. Instn Civ. Engrs, Civ. Engng, Channel Tunnel: Part 1—Tunnels*, 1992, 103–126.

6. FUGEMAN I. C. D. *et al.* Major undergrund structures. *Proc. Instn Civ. Engrs, Civ. Engng, Channel Tunnel: Part 1—Tunnels*, 1992, 87–102.

7. FUGEMAN I.C.D. *et al.* The Channel Tunnel: development of design and construction methods for the United Kingdom crossover. *Proc. Int. Symp. Tunnelling '91*, London, 1991.

8. CRAWLEY J. D. and POLLARD C. Ground treatment to improve tunnel progress on the Channel Tunnel marine drives. *Ground Engineering*, 1992, Jan.–Feb.

9. POLLARD C. *et al.* Survey and alignment control for the Channel Tunnel—United Kingdom drives. *Proc. Int. Symp. Tunnelling '91*, London, 1991.

Major underground structures

*I. C. D. Fugeman, BEng, MICE, J. Hawley, MA, MICE, MASCE, and
A. G. Myers, BEng*

*Proc. Instn Civ.
Engrs, Civ. Engng,
Channel Tunnel
Part 1: Tunnels
1992, 87–102*

Paper 9933

**Underground structures other than the
machine-driven tunnels account for a significant proportion of UK tunnelling
works. The construction techniques for
these structures can be separated into two
categories: New Austrian Tunnelling
Method (NATM) tunnels where primary
support is followed by an in situ concrete
permanent lining; and traditional hand-
excavated structures using segmental
linings. This Paper considers in detail the
three major underground structures:
Shakespeare underground development,
Castle Hill tunnels and the UK undersea
crossover.**

Introduction

General resume

The various underground structures form a significant proportion of the Channel Tunnel
project. The type of tunnel boring machine
(TBM) and the choice of an expanded concrete
lining of a particular thickness were key decisions that had to be taken at a very early stage
in the project in 1986–87. However, it was the
function of the underground structures evolving from later multidisciplinary development
studies that determined their final form. The
timing of the detailed design of the construction of each underground structure was such
that each element benefited from experience
gained on the previous structure. The development of the use of the New Austrian Tunnelling
Method (NATM) on the Channel Tunnel project
over a four year period was a significant mile-
stone in its application in British engineering.

2. This Paper addresses the relationship
between construction methods and programme
and the functional requirements and form of the
underground structures.

3. The application of NATM techniques
requires a strong management structure with
extensive yet precise procedures. These are
also the key features of a quality-assured
system. This discipline provides the correct
environment for safe construction, an aspect of
paramount importance in underground construction (see Table 1).

New Austrian Tunnelling Method

4. The tunnelling techniques employed for
the underground structure were first systematically described and implemented by Austrian
engineers in the 1960s. The advantages of this
method are speed of mobilization, freedom to
vary cross-section and ease of forming junc-

tions. These were particularly marked on the
Channel Tunnel project, first with the complex
of tunnels beneath Shakespeare Cliff and subsequently at the UK undersea crossover (Fig. 1).

5. The essence of the NATM as applied to
the Channel Tunnel was control over the redistribution of stress in the ground as the excavation advanced, by adjusting the construction
sequence, the primary support and the timing
of its installation. Design calculations, particularly for the primary lining, needed to take
account of the construction sequence and
timing, and this was an essential aspect of the
methods developed for the analyses. It should
be stressed that the application of the NATM
was controlled by a written procedure which
governed all changes to the design or to the
construction method or sequence. It included
the procedure to be followed in the case of an
unpredicted event affecting the stability of the
excavation or structure, such as greater than
expected deformations or primary lining
failure.

6. Excavation was by roadheader, with an
advance of $1\frac{1}{2}$ to 2 m. A lightweight lattice arch
was erected close to the face, a layer of shotcrete sprayed onto the excavated periphery,
and rock dowels installed to a predetermined
pattern. Depending on the size of the tunnel and
the ground conditions, a second layer of shotcrete reinforced with steel mesh and additional
rock dowels was installed. The performance of
the primary support was monitored routinely
by measuring convergence and settlement
within the tunnel, and at some locations by
more extensive instrumentation measuring
ground loads on the primary lining, stresses
within the lining, groundwater pressures, and
movements within the ground around the
tunnel. Although the measurements usually
confirmed that the support was appropriate, in
some cases the support was modified or additional support was applied.

7. A permanent secondary lining of in situ
concrete was later installed to carry long-term
loads arising from the build-up of water pressure, creep of the ground, and to allow for any
deterioration of the primary support.

Hand tunnelling

8. Some structures were excavated using
soft ground tunnelling techniques by tunnel
miners using hand-held tools. Temporary face
support was provided by traditional face tim-
bering. The bolted segmental tunnel linings
were manufactured from spheroidal graphite
iron (SGI) and were erected by hand or with the

*I. C. D. Fugeman,
Engineering
Manager—Design,
Transmanche-
Link*

*J. Hawley,
Principal
Engineer,
Tunnels Division,
Mott MacDonald*

*A. G. Myers,
Section Manager,
NATM and Hand
Tunnels,
Transmanche-
Link*

Table 1. Data sheet UK tunnelling works—excavated volumes

	Length: m	Quantity: m³/m	Excavated volumes: m³
TBM driven tunnels			
Running tunnel, sea	36 608 m	54·4 m³/m	1 991 475
Running tunnel, land	15 712 m	59·2 m³/m	930 150
Service tunnel, sea	21 803 m	22·4 m³/m	488 387
Service tunnel, land	7926 m	25·73 m³/m	203 936
Total TBM driven tunnels	82 049 m		3 613 948
NATM	Average length on centre line of running tunnel		
Shakespeare underground development	516 m		150 000
UK undersea crossover	164 m		48 000
Castle Hill	532 m		88 000
Total NATM			286 000
Hand Tunnelling	Total no.		
Pumping stations (including NATM)	2		22 300
Land Sea			
Cross-passages 22 58	80		17 600
Technical rooms 20 39	59		13 900
Pressure relief ducts 32 74	106		10 600
Total hand tunnelling			64 600
Total excavation			3 964 348

aid of a winch. Grout was injected behind the linings to fill the annulus.

9. The techniques were conventional and basic, and could be adapted for the very worst ground conditions. For these reasons hand methods were adopted for all the repetitive smaller structures which had to be constructed intermittently and reliably throughout the full length of the project.

10. Each workstation was set up to be self-sufficient, and included a substation with transformer capacity and switchgear to run its own compressor. Up to four workstations were in operation in each of the three tunnels at any one time.

Design principles

11. Before commencing detailed design of the underground structures, design criteria and design data were derived and assembled into design manuals.

Design criteria

12. There are no British or international codes relating specifically to the structural design of tunnels. However, BS 8110: Structural use of concrete,[1] was followed where applicable, with reference also being made to the German code for reinforced concrete structures, DIN 1045.[2] The design was governed by loadings due to ground and groundwater pressures, with the self-weight of the structures considered as a secondary effect. The following load and material factors were adopted for the ultimate limit state.

(a) Load factors
    (i) permanent gravity loads (ground and water): best estimate values of geotechnical parameters 1·35; worst credible values of geotechnical parameters 1·2
    (ii) variable and imposed loads 1·5
    (iii) accidental loads 1·05
    (iv) construction loads 1·2
(b) material factors for normal load combinations

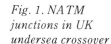

Fig. 1. NATM junctions in UK undersea crossover

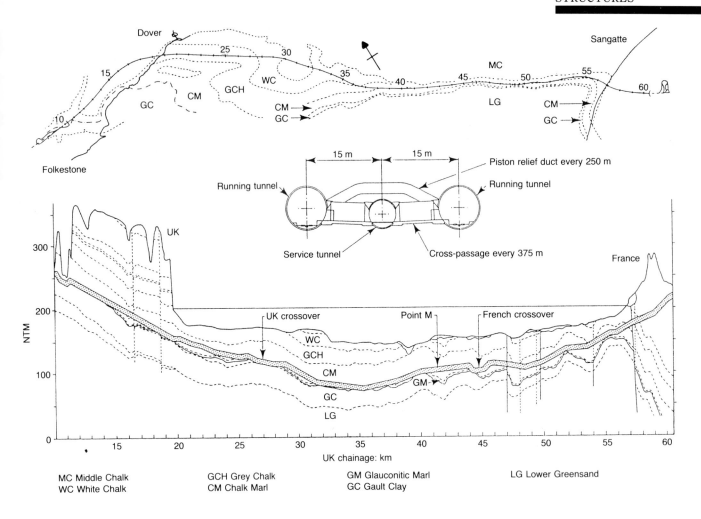

*Fig. 2. Channel
Tunnel geology*

MC Middle Chalk   GCH Grey Chalk   GM Glauconitic Marl   LG Lower Greensand
WC White Chalk   CM Chalk Marl   GC Gault Clay

(i) concrete in compression 1·5
(ii) steel reinforcement 1·15
(c) material factors for exceptional or accidental load combinations
(i) concrete in compression 1·3
(ii) steel reinforcement 1·0

13. The permanent linings also had to be designed for the effects of an earthquake producing up to 0·05 g horizontal acceleration. This was not a critical case, because of the reduced load and material factors associated with exceptional load combinations. Junctions between dissimilar structures or cross-sections were detailed to allow relative movement of the magnitude, typically around 5 mm, expected during the design earthquake.

14. The effects of other loadings, such as those due to excavation equipment, TBM forces, wedge failure and local effects on elements of the support system, were also checked and taken into account where relevant. In some cases, for example where the running tunnels were driven close to previously excavated chambers at the pumping stations, a limit was imposed on tunnel boring machine gripper forces.

15. It was a contractual requirement that the permanent lining could withstand the direct application of the full unfactored ground overburden and water pressure. For non-circular cross-sections this was interpreted by taking the largest dimension as the diameter of an equivalent circular tunnel and applying the load to cause a uniform radial pressure. This loading was checked and found to be critical for many of the permanent linings, though not for the crossover cavern.

Design data

16. The geology of the UK section of the Channel Tunnel is summarized in Fig. 2. Undersea, along the tunnel route, the Chalk lies immediately beneath the superficial seabed deposits. It is divided into Grey Chalk, which is weathered to varying degrees, below which is the Chalk Marl. The Chalk Marl is underlain by 2–3 m of Glauconitic Marl, a marly sandstone with characteristic green flecks. Beneath the Glauconitic Marl lies a transition zone designated Zone 6A, of variable thickness up to 8 m, extending to the Gault Clay. The optimum tunnelling horizon lies in the unweathered Chalk Marl, generally 5 m above the base of the Glauconitic Marl.

17. Values of geotechnical parameters were derived from the results of site investigation,

and representative geotechnical parameters are listed in Table 2. The following properties of the ground were taken into account by the analyses.

18. Rock mass strengths were estimated from the intact effective strength values obtained from laboratory testing, and were expressed in terms of the effective cohesion intercept C' and the angle of shearing resistance ϕ'. After yield, the angle of shearing resistance was assumed unchanged, but the cohesion intercept was reduced. For the undersea crossover, composite shear strength parameters were estimated using the Hoek and Brown[3] failure criterion, and they are listed in Table 2 as C'_1 and ϕ'_1 for values of the minor principal stress less than about 500 kN/m^2, and C'_2 and ϕ'_2 for greater values of the minor principal stress.

19. Rock mass stiffnesses E_u were initially based on the intact rock undrained modulus at 50% of failure stress. Monitoring of the early excavations at Shakespeare underground development showed these values to be too low, presumably because strains were very small except locally to the excavation. For later calculations, rock mass stiffnesses were based on the undrained modulus from unload–reload tests, divided by an appropriate empirical factor

derived from measurements during the earlier excavations.

20. Laboratory tests and monitoring of the 1975 works showed that the ground exhibits creep. Creep factors were assigned to each stratum, defined so that for imposition of uniaxial stress

$$\varepsilon_t = (1 + \theta_t)\varepsilon_0$$

where ε_0 is the initial strain, ε_t is the strain at time t, and θ_t is the creep factor corresponding to time t. Values of the creep factor at 90 days and 1 year were used in calculating the primary lining behaviour at stages in construction and prior to installation of the permanent lining, and were taken as 45 and 75%, respectively, of θ_{120} years.

21. The primary lining shotcrete has a specified characteristic strength of 25 MN/m^2, as measured by 100 mm cylinder tests at 28 days. A factor of 0·85 was used, according to DIN 1045,[2] to derive the uniaxial compressive strength of 21·25 MN/m^2. It was assumed that the shotcrete would eventually degrade to a granular material with a friction angle of 30°, shedding load onto the permanent linings.

22. In order to comply with the durability criteria established for the Channel Tunnel project, the in situ concrete for undersea struc-

Table 2. Representative values of geotechnical parameters

	Upper Chalk Marl	Lower Chalk Marl	Basal Chalk Marl	Glauconitic Marl	Gault Clay
Shakespeare underground development					
Intact rock strength C_u : kN/m^2	4000	2500	—	2350	1500
Rock mass peak strength C' : kN/m^2	200	900	—	500	250
ϕ' : deg	40	35	—	30	30
Rock mass stiffness E_u : MN/m^2	1050	950	—	950	350
Creep factor θ at 120 years	1·2	1·3	—	1·3	2·0
*Castle Hill tunnels**					
C_u : kN/m^2	4000	2625	—	2350	1500
	(2000)	(2000)	—	(2000)	(1000)
C' : kN/m^2 (Peak)	200	900	—	500	250
ϕ' : deg	40	35	—	30	30
E_u : MN/m^2	750	600	—	600	175
	(60)	(60)	—	(60)	(40)
θ	1·5	1·5	—	1·5	2·1
	(2·1)	(2·1)	—	(2·1)	(2·1)
UK undersea crossover (best estimate values)					
C_u : kN/m^2	4250	1750	3750	6500	1000
C'_1 : kN/m^2	250	135	225	275	70
ϕ'_1 : deg	39	35	40	43	32
C'_2 : kN/m^2	500	350	480	575	275
ϕ'_2 : deg	28	24	29	34	19
E_u : MN/m^2	825	400	750	1000	300
θ	1·0	1·0	1·0	1·0	2·1

*Data for tunnels east of landslip zone; data for tunnels within landslip zone in parentheses.

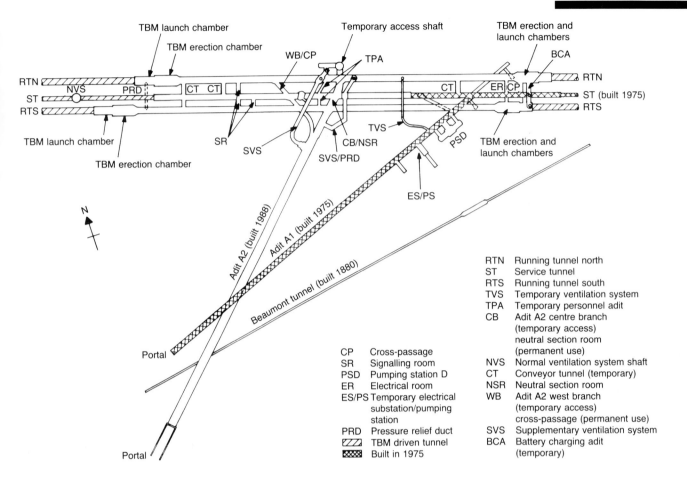

RTN   Running tunnel north
ST   Service tunnel
RTS   Running tunnel south
TVS   Temporary ventilation system
TPA   Temporary personnel adit
CB   Adit A2 centre branch
       (temporary access)
       neutral section room
       (permanent use)
NVS   Normal ventilation system shaft
CT   Conveyor tunnel (temporary)
NSR   Neutral section room
WB   Adit A2 west branch
       (temporary access)
       cross-passage (permanent use)
SVS   Supplementary ventilation system
BCA   Battery charging adit
       (temporary)

CP   Cross-passage
SR   Signalling room
PSD   Pumping station D
ER   Electrical room
ES/PS Temporary electrical
       substation/pumping
       station
PRD   Pressure relief duct
▨▨   TBM driven tunnel
▩▩   Built in 1975

*Fig. 3. Shakespeare
underground
development*

tures has a minimum characteristic strength of
40 MN/m² at 28 days.

Design methods

23.   Most of the design was verified by two-
dimensional analysis using finite-element computer programs. The ground was taken as
linear-elastic up to failure, and perfectly plastic
with a reduced post-peak strength after failure.
Depending on the particular computer program
used, either the modified Drucker–Prager
criterion[4] was used to model failure of the rock
mass, or the Hoek and Brown[3] criterion was
used directly. The numerical model chosen for
design simulated the participation of the rock
mass in the load-bearing process and the three-
dimensional effects as follows.

24.   A first calculation step modelled the
stress redistribution due to the excavation
process up to the time of installing the shotcrete lining. A second calculation step took
account of the lining and the further stress
redistribution up to the time when the excavation process no longer influenced the investigated cross-section. A third calculation step
took account of the time-dependent behaviour
of both the shotcrete and the rock mass.

25.   The individual construction stages were
modelled in the finite-element analyses. The
behaviour and effect of young shotcrete near

the tunnel face in the second calculation step
was modelled by assigning a hypothetical
modulus of elasticity to the shotcrete as
described by Pöttler.[5] This modulus is less than
the true modulus, to take into account the
stiffness history of the shotcrete, its creep
behaviour and the time-dependent loading due
to the excavation process and advance of the
tunnel face.

26.   In some cases, for example in
developing standardized design procedures for
the many junctions in the Shakespeare underground development and for analysing some
cross-passage junction structures, full three-
dimensional finite-element analyses were performed, but only elastic behaviour was
modelled.

27.   Rock dowels and lattice girders were not
modelled or explicitly taken into account in the
finite-element analyses. The effect of dowels is
implicit in the values adopted for the strength
and stiffness of the rock mass. The stiffness
and strength of a lattice girder is low compared
with that of hardened shotcrete, but the load
that a lattice girder can take was investigated
in separate calculations.

28.   Predictions of ground load, lining stress
and deformation were made before commencing
construction. The predictions of deformation
made allowance for movements that would

occur before instruments could be installed. Although measurements gave a wide scatter, particularly for loads and stresses, construction experience confirmed the design methods and data employed. It is impracticable to give any meaningful summary of the mass of measurement results collected, but an example and ranges of certain readings are given later in this Paper for the crossover. There was no requirement arising from NATM procedures to monitor the permanent linings, though some permanent instrumentation was installed as part of long-term tunnel instrumentation. There was a contractual requirement to monitor the behaviour of the linings and ground conditions throughout the Fixed Link.

Shakespeare underground development (SUD)

Function

29. The rapid construction of the Shakespeare underground development was a critical requirement for the succcess of the Channel Tunnel, because six TBMs were to be erected and launched from this one location. This entailed the excavation of high chambers, plus a system of smaller tunnels associated with the marshalling and spoil disposal facilities (as described by Pollard[6]); the choice of the NATM system for these works in late 1986 was without precedent in British tunnelling projects. Rapid advance and flexibility for additional construction facilities amply justified this bold decision and provided the experience for the application of NATM in other areas of the works.

30. After three years of meeting the temporary construction requirements for access of construction materials and egress of excavated spoil, the works were adapted to serve as the delivery route for the permanent supplementary ventilation system and other serivces from the cooling plant buildings which form part of the permanent Shakespeare facility.

31. Key quantities of the Shakespeare underground development are as follows. NATM tunnels 2720 m; junctions 65 No.; excavation 160 000 m³; shotcrete area 40 000 m². Adit A2: excavated area 70 m²; width 12·1 m; height 7·3 m. Typical RT erection chamber: excavated area 150 m²; length 25 m; width 13·4 m; height 15 m.

Construction

32. Tunnelling work commenced at Shakespeare Cliff on three fronts, as follows.

33. Hand tunnelling methods were used to break out from the existing 490 m long 1975 access adit A1 and commence the construction of the conveyor tunnels. These comprised the warren of passages required underneath the three main tunnels to house conveyors for transferring spoil from underground tipping bunkers to the spoil conveyor running to the surface (Fig. 3). These tunnels were lined with 3000 mm ID and 4000 mm ID SGI rings with mass concrete at the junctions. The tunnels were excavated through Gault Clay, and progress rates of 12 rings per week were achieved in the 4000 mm ID main passage. Excavation and ring building was by conventional hand methods with a four-man mining gang.

34. Shaft sinking started from the upper site for an access shaft of internal diameter 9250 mm and 110 m deep. The initial 10 m was segmentally lined, and the remainder was constructed using colliery arches spaced at 1200 mm centres with 150 mm thick temporary lining shotcrete sprayed on to the wall surface. Shaft sinking was completed in approximately 5 months. A permanent in situ concrete lining was slipformed once the existing underground works had been connected to the base of the shaft.

35. The new main access adit A2 on the Lower Shakespeare Site (Fig. 3) was originally envisaged as a deep open cut. However, it was decided to construct the works in tunnel from very close to the surface utilizing NATM techniques. The first 80 m of tunnel was through loose chalk fill which overlaid the original beach sand, and these allowed tidal water flow. The size of the adit A2 was determined by the requirement to install five rail tracks which resulted in a width of 12·5 m and a height of 7 m. In the loose fill, injected steel pipes were installed as spiling prior to excavation, which was then carried out by a special Atlas HD 1604 360° excavator or by hand. In broken ground, progress was restricted to 10 m/week. When virgin Upper Chalk was encountered, water had to be contended with through the chalk fissures. Once the Chalk Marl was reached, the invert was flattened and progress increased to a maximum 35 m per week. Early invert closure became necessary at the bifurcation point in the tunnel where the width increased to 13·5 m and the Gault Clay horizon was only 1 m below the invert. A Demag H55 roadheader was used for the main adit A2 excavation with a CAT 966 loading into Moxy dump trucks. A purpose-made twin shotcrete rig was used and rock bolting was carried out by a SIG twin boom electro-hydraulic drill rig with basket. A Poclain P61 excavator was specially adapted as a man hoist platform for shotcreting and for lattice arch and mesh erection. Once the access shaft had been sunk a Voest Alpine AMT 70 roadheader was lowered down and tunnelling for the land service tunnel marshalling area and erection chambers commenced.

36. One of the running tunnel (RT) erection chambers, 13·5 m wide, 14·5 m high and 45 m

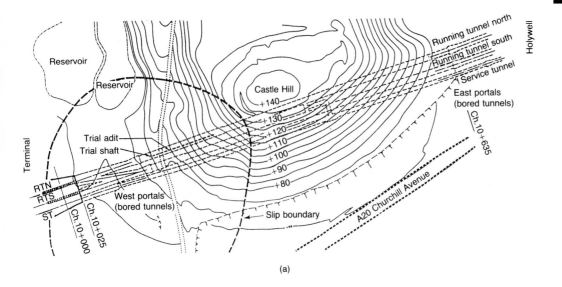

(a)

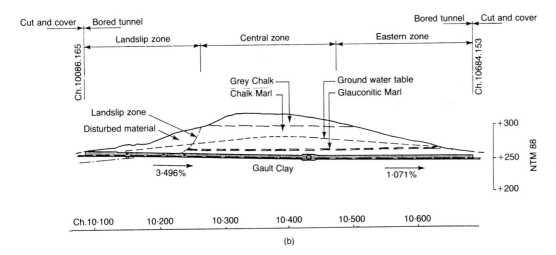

(b)

*Fig. 4. Castle Hill:
(a) tunnels and
landslip;
(b) geotechnical
sequence zones*

long was completed in just 30 days. The maximum advance in one week in the running tunnel marshalling tunnels was 50 m.

37.   The construction of the main SUD complex proceeded with the two main road-headers (Demag H55 and Voest Alpine AMT70) employed on three working faces. Access to these faces was required during the erection of the TBMs and their backup trains. Initially spoil disposal from the excavation of the service tunnel backshunt and the land service tunnel occurred via the shaft. After the commencement of the TBM drives the excavation of SUD continued, sharing a common access route via adit A2. This entailed the rubber-tyred dump trucks hauling spoil up adit A2 adjacent to the rail tracks feeding materials down to the TBM drives. Muck skips from the service tunnel marine drive initially exited via adit A1 and then via adit A2 rack track, before the completion of the extensive underground bunker system and the commissioning of a conveyor system in adit A1.

Castle Hill tunnels

Choice of NATM and design aspects

38.   The alignment and profile of the 500 m long Castle Hill tunnels was dictated by gradient limitations, alignment within the adjacent Holywell Coombe to the east (which in turn was dictated by a site of special scientific interest) and the level of the UK terminal to the west. The service tunnel moves away from its standard position between the two running tunnels to exit south of the running tunnels at the portal. At the Castle Hill portal the running tunnels were enlarged to incorporate a cross-over.

39.   The tunnels are located in mixed faces of Gault Clay, Glauconitic Marl and Chalk Marl (Fig. 4) and pass through the disturbed ground of an ancient landslip zone. This feature required extensive stabilization works in the form of 50 000 m³ of toe weighting, under drainage, and the installation of an extensive long-term monitoring system. Key factors

Fig. 5. Castle Hill portal

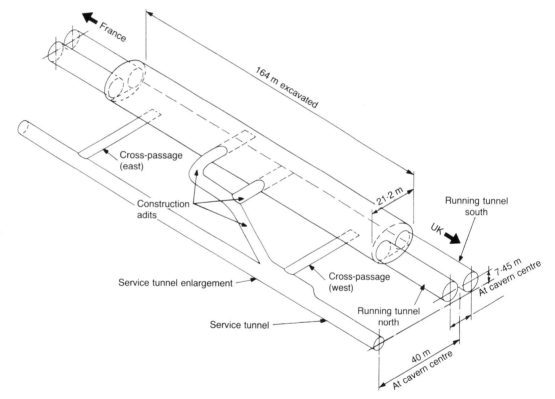

Fig. 6. UK undersea crossover—isometric

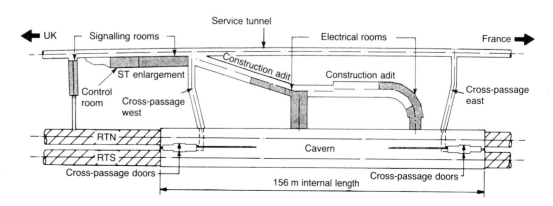

Fig. 7. UK undersea crossover—technical room layout

which contributed to the choice of NATM were

(a) Flexibility of the method to cope with variable and unpredictable ground conditions;

(b) Ease of construction of enlarged cross-sections through the landslip zone, and formation of junctions for drainage galleries;

(c) availability of tunnelling equipment and materials to meet the construction programme, which did not permit waiting for the arrival of the landward TBM drives;

(d) lower construction cost.

40. A trial heading had shown that the ground was blocky, and analysis of the effects for the full-sized tunnels indicated that fore-poling ahead of the face and immediate application of support would be necessary in the worst areas.

41. In the landslip area the running tunnel cross-section was widened by 630 mm so that the required structure gauge could be maintained in case of any lining deformation or lateral shift. Transverse movement joints were provided, and the lining was reinforced near the slip failure plane.

Construction

42. All three tunnels were constructed from east to west, commencing with the service tunnel. Forepoling using 4 m long spiles was a feature from the start, but the ground conditions experienced were a severe test of the flexibility of the NATM system. In the Gault Clay the cohesion and friction along the joint surfaces were very low. Face stability became a major problem; this was initially resolved by a reduction of the height of the face, and ultimately by the use of a 2 m rake on the lattice arches to allow a sloping buttressed face. The running tunnel excavation using a heading and bench method was in slightly more favourable ground conditions, but nevertheless it did require amendment to the excavation sequence before a satisfactory sequence of working was established. Deformations were higher than expected, and back-analyses using lower rock mass stiffnesses gave good agreement with the measurement results and indicated the need for an increase in the thickness of the shotcrete primary lining.

43. Water ingress was not a problem during excavation and shotcreting, but it gave difficult working conditions when mixed with the invert clay. This was exacerbated by the plant movements, especially in the service tunnel where the AMT 70 road header was not the ideal choice although perfectly suited to the running tunnel profile. This machine was too large to permit the passage of other equipment to the face of the service tunnel and had to be tracked out after each advance until a passing bay was constructed at the centre of the tunnel. Two other key items of plant were an adapted

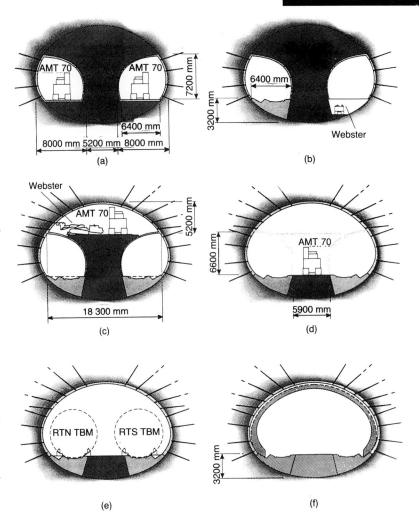

Tamrock four-wheel-drive articulated drill rig fitted with one boom and a telescopic basket, and a remote-controlled hydraulic spraying arm attached to a four-wheel-drive chassis. The latter was an improvement on the manual spraying techniques used at Shakespeare. Shotcrete was site-batched above ground at Holywell and delivered to the face by conventional truck mixers. Muck was carried out from the face by Moxy dump trucks. A more detailed account of the development of these tunnels has been given by Penny *et al.*[7]

44. Although the NATM was utilized successfully for the Castle Hill tunnels, it did demonstrate the need for careful planning, and an organization and management structure capable of rapidly dealing with design modifications required to match changes in working method. A notable element of the Castle Hill tunnels was the portal structure (Fig. 5). A major architectural input was necessary to produce a gentle curved structure which both satisfied the requirements of the Royal Fine Arts Commission and also matched the importance of being Britain's gateway to Europe.

Fig. 8. UK undersea crossover—cavern construction sequence: (a) sidewall drift heading; (b) sidewall drift invert construction; (c) crown heading; (d) shotcrete removal and bench excavation; (e) RTN and RTS TBM passing; (f) invert concrete and in situ secondary lining

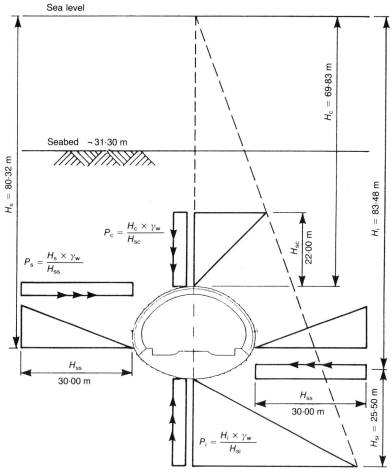

Initial rock stresses:

Vertical $P_v = \gamma_{uw} \times H_G$

Horizontal $P_h = \gamma_{uw} \times H_G \times K_0$

H_G Depth below seabed

K_0 Coefficient of lateral pressure

γ_w Density of water

γ_{uw} Bouyant density of rock mass

$P_i P_s P_c$ Water seepage forces acting on rock mass

$H_{sc} H_{ss} H_{si}$ Distances vary depending on geotechnical parameters, typical values shown

Fig. 9. UK undersea crossover—load case 3

UK undersea crossover

Function

45. Two undersea crossovers connecting the running tunnels are required to achieve maximum throughput during single-line working. This occurs almost nightly when sections of the tunnel will be isolated to gain access for maintenance and during breakdown or emergency evacuation scenarios. Theoretically the maximum throughput occurs if the crossovers are located at third points between the rail crossing facilities at the UK and French portals, as described by Fugeman *et al.*[8]

46. By mid-1987 the initial joint UK/French studies had concluded that a diamond crossing contained within a single structure could be constructed in the prevailing ground condi-

tions, could meet the programme constraints, and could provide the desired operational features. UK and French schemes were then developed independently based on a common track geometry and performance requirements (Fig. 6). In the UK crossover the electrical, control and signalling rooms are accommodated within the access adits (Fig. 7) outside the cavern structure, and in the French cavern they are located above the track. Both schemes utilized a common design for the large longitudinal sliding doors which close across the diamond to achieve total segregation between the two running tunnels during normal two-track operation.

47. The UK marine service tunnel drive had the advantage of an early start in comparison with the other UK and French tunnels due to the existing 1975 access adit A1 and a short section of service tunnel. This created a theoretical 9 month window between the service tunnel TBM passing the location of the UK undersea crossover and the arrival of the running tunnel TBMs. It also allowed a $2\frac{1}{2}$ year period for development of the crossover scheme. By comparison, the French had a theoretical window of 2 months which could easily have been eroded during 2 years of tunnelling over 15 km length, but had a development period of $3\frac{1}{2}$ years before commencing construction. Differing geology and different time windows between the service tunnel and the running tunnel meant that totally different methods of construction were required for these large undersea structures.

Design

48. Its size and undersea location meant that the design of the crossover cavern marked a significant step beyond what had been proved by experience at the time. The design took account of the construction sequence shown in Fig. 8 and the construction programme of Fig. 11, and a range of calculations were performed using

(a) best estimate and worst credible values of geotechnical parameters;

(b) creep factor and double Kelvin cell models for creep;

(c) a range of load assumptions;

(d) a range of values for the hypothetical modulus of elasticity.

49. Three load cases were selected to represent a range of groundwater pressure distributions

50. In load case 1, water pressure was applied to the seabed and the initial effective stresses in the rock were based on its saturated density. Although the provision of back-drainage and the adjacent permeable segmentally lined tunnels made it unlikely that full hydrostatic head could ever develop, load case 2 assumed full water head on the permanent

lining as the greatest credible water load that
could apply, in conjunction with initial effective stresses in the rock mass corresponding to
its submerged density. Load cases 1 and 2 represent the extremes of the possible water pressure distributions. Load case 3 (Fig. 9) modelled
water pressure increasing with depth except
within a certain distance of the cavern, from
where it decreased linearly to the excavation
boundary. Note that the expression for seepage
force P_h (and P_1) in Fig. 9 corresponds to the
pressure gradient plus and minus γ_w to take
account of the weight of water within the
affected zone. The distribution of water pressure was based on the results of flow net
analyses.

51.   In addition, wedge analyses were performed to confirm the pattern of rock dowels
chosen, and face stability was checked for the
crown heading. The design of the shotcrete
lining of the sidewall drifts included loadings
from the construction equipment for crown
heading excavation.

52.   By the time the detailed design of the
crossover commenced in November 1988, tunnelling on the project was well under way, with
excavation of the access and marshalling
tunnels at Shakespeare essentially complete,
and the TBM drives well advanced. This gave
much valuable information about the ground
response to a range of tunnelling techniques,
though the service tunnel drive had not yet
reached the crossover location.

53.   Following the phase 2 marine site investigation undertaken in 1988, which included
three additional boreholes and improved the
accuracy of strata levels, the crossover location
was adjusted to avoid areas of minor faulting
identified in the chalk.[9] Its level was fixed to
give about 5 m of zone 6A material beneath the
invert and 36 m cover between the top of the
excavation and the seabed.

54.   Both best estimate and worst credible
values of geotechnical parameters were derived
from the results of the various phases of site
investigation. The best estimate values were
conservatively chosen to given calculation
results which would be comparable with actual
behaviour, and calculations using the worst
credible values gave an assurance of stability
for a range of possible ground conditions.

55.   The numerical models used for the
design have been described by Hawley and
Pötter,[10] and included an analysis in which the
three most persistent joint directions were modelled explicitly, these being sub-horizontal
bedding, shallow-dipping joints at 10–30°, and
subvertical jointing, the latter two both striking
roughly parallel to the cavern axis.

56.   Some of the predictions of the results of
measurements of ground and primary support
behaviour made before commencing construction are given in Table 3, together with the

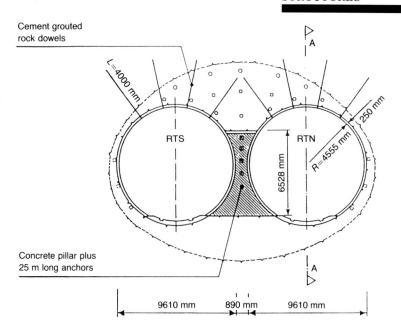

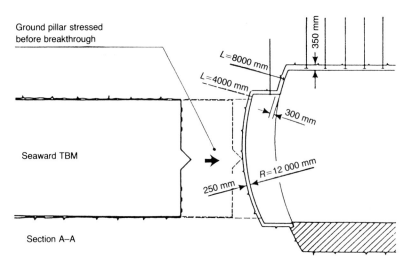

Fig. 10. UK undersea crossover—additional support measures
prior to breakthrough of running tunnel north TBM

Table 3. Predicted and actual results of geotechnical measurements

	Calculated range	Measured range*
Cavern roof settlement: mm		
Initial values	17–31	8–41
Values after 1 year	18–37	22–66
Cavern shotcrete stresses: MN/m²		
Crown	5–7	2–12
Initial		
1 year		
	7–11	5–11
Sidewalls	8–13	1–4
Initial		
1 year		
	13–17	3–10

* Not including unpredicted event.

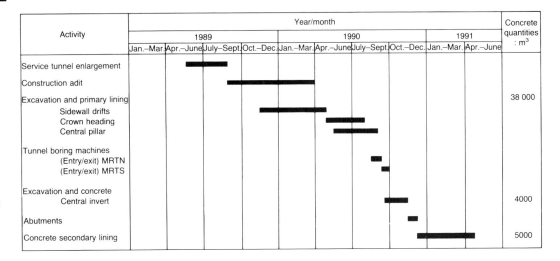

Activity	Year/month									Concrete quantities : m³
	1989				1990				1991	
	Jan.–Mar.	Apr.–June	July–Sept.	Oct.–Dec.	Jan.–Mar.	Apr.–June	July–Sept.	Oct.–Dec.	Jan.–Mar.	Apr.–June	
Service tunnel enlargement		▬	▬								
Construction adit			▬	▬							
Excavation and primary lining											38 000
Sidewall drifts				▬	▬						
Crown heading						▬					
Central pillar						▬					
Tunnel boring machines											
(Entry/exit) MRTN							▬				
(Entry/exit) MRTS							▬				
Excavation and concrete											4000
Central invert								▬			
Abutments									▬		
Concrete secondary lining									▬	▬	5000

Fig. 11. UK undersea crossover construction programme

range of actual measured results.

57. For the stability analysis of the permanent lining, the calculations of the primary lining were continued using the same numerical models. The permanent lining in the cavern consists of unreinforced in situ concrete with a nominal minimum thickness of 600 mm. The permanent lining is waterproofed by a geotextile backed membrane system.

58. At the ends of the cavern 2 m long tunnel eyes were formed for the reception and departure of the running tunnel TBMs (Fig. 10). Beyond the cavern, the rock pillar between the running tunnels as they approached the crossover narrowed to little over 2 m in width, and this was analysed using a finite-difference program FLAC[11] that modelled the joint sets in the ground and individual elements of possible support measures. The calculations resulted in

constraints on the construction sequence for the running tunnels over lengths of 150 m at either end of the crossover, with use of bolted SGI lining, dowelling and propping in the first (north) running tunnel for 114 m up to the cavern. SGI is required for 21 m either side of the cavern to provide additional width to match the increased structure gauge associated with the track turnouts for the crossover. The toe of the track turnout switch occurs 7 m into the driven tunnels either side of the cavern. This feature results in the minimum length for the excavated cavern.

Construction

59. The construction sequence is best described by reference to the programme (Fig. 11) and the service tunnel enlargement and construction adit layouts (Figs 12 and 13). Con-

Fig. 12. UK undersea crossover service tunnel enlargement—construction sequence: (a) plan of ST enlargement; (b) enlarged section A-A during construction; (c) enlarged section A-A after completion

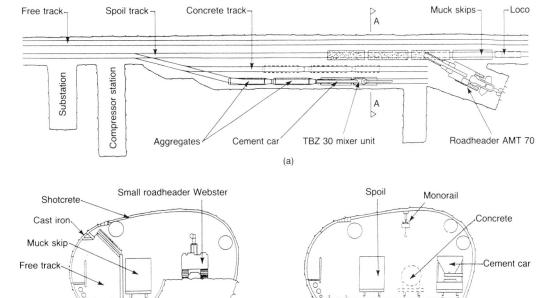

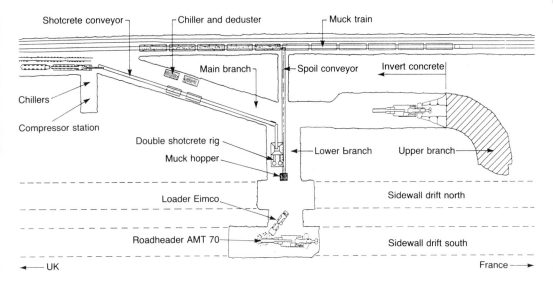

Fig. 13. UK undersea crossover construction adits—construction sequence

struction of the cavern commenced with sidewall drifts followed by the crown heading, bench and invert (Figs 8 and 14). This sequence, instead of a heading/bench/invert method, was chosen primarily for safety reasons and also because

(a) sidewall drifts serve as both pilot and drainage tunnels;
(b) face stability is first established on a small scale;
(c) the sidewall drift walls provide a secure foundation for the top heading arch.

60. The original concept was for the crossover to be built from the service tunnel within a 9 month window before the arrival of the first running tunnel TBM. In reality the construction rate of the service tunnel improved at the time that both running tunnels were experiencing some delays. The window increased to 16

months but the crossover works actually suffered as the lead distance of the ST TBM over the RT TBM lengthened. This increased lead imposed greater logistical constraint and reduced the efficiency of the ventilation system. Access along the service tunnel was effectively limited to a single line of track owing to the construction of cross-passages, technical rooms, a main pumping station and 700 m of ground treatment in advance of each of the running tunnel TBMs. Access and material supply problems greatly delayed the initial service tunnel enlargement and the sloping construction adits.

61. The main ventilation system for the cavern incorporating 4 dedusters could not be installed until the completion of the construction adits. A cooling system utilized seepage water collected from sumps 5 km downstream

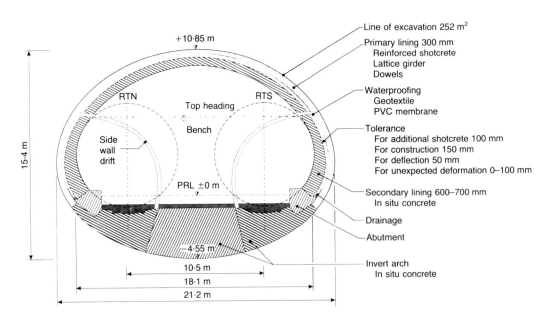

Fig. 14. UK undersea crossover cavern cross-section

of the crossover towards France. Two chiller units and heat exchangers were established, but not until work had commenced in the crown heading. Until that time work in the sidewall drifts and its associated large concrete invert pours created undesirable working conditions with high temperatures and humidity.

62. Geotechnical measurements form an integral part of the NATM system and are carried out to determine rock mass behaviour until a state of equilibrium is obtained. Measurement stations were located at predetermined positions, typically at 15 m centres, with additional stations at tunnel junctions. Instruments and deformation reference points were installed with the primary support within 0·5–2·0 m of the face at each construction stage.

Typical instrumentation for the cavern and an example of deformation results are shown in Fig. 15, indicating the rate of movement decreasing over about 15 days to give a stabilized initial value at a distance of 25–30 m (about 1½ tunnel diameters) behind the face. The ranges of some measured initial instrumentation results are compared with calculated values in Table 4.

63. After excavation of the sidewall drifts had been completed, and towards the landward end of the crown heading excavation, an unpredicted event occurred. Starting about 6 m behind the face over a length of 16 m, cracks appeared in the 250 mm thick mesh-reinforced shotcrete along both sides of the crown heading, and the inner bars of the lattice

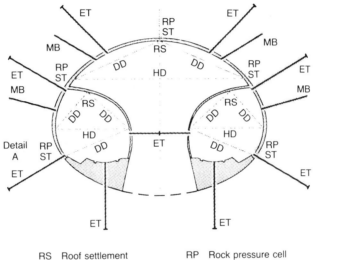

Cavern contained 200 instrumentation stations, requiring 120 readings daily

Detail A

RS Roof settlement
SS Sidewall settlement
HD Horizontal deformation
DD Diagonal deformation
FH Floor heave

RP Rock pressure cell
ST Shotcrete pressure cell
ET Extensometer
MB Measuring rock bolt

(a)

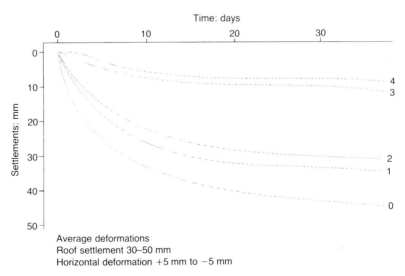

Average deformations
Roof settlement 30–50 mm
Horizontal deformation +5 mm to −5 mm

Fig. 15. UK undersea crossover geotechnical measurement: (a) typical cavern measurement section; (b) typical settlements

(b)

girders were observed to have buckled at the cracks. This event occurred over a period of 3 hours and movements then ceased. Measurements showed the crown settlement to have increased from around 25 mm to around 60 mm. The shotcrete hoop stresses at a measuring section within the affected area 19 m behind the face were reduced from about 10 MN/m² measured two days before the event down to 0–7 MN/m². To cause the observed compression failure, stresses must have exceeded the shotcrete strength of around 18 MN/m² at its age of 4 days.

64. Two main factors were identified as contributing to the event. The heading face had just passed a transverse discontinuity in the rock, interrupting the three-dimensional distribution of loads, and putting higher loads than normal on the young shotcrete. Rock dowel holes drilled after the event found significant amounts of water, suggesting that high water pressures were acting on a relatively impermeable layer located only about 4 m above the crown.

65. The second of these factors represented a potentially more critical load case than those derived from the known ground conditions and considered during design. Additional calculations were carried out to study the influence of the individual factors thought to have contributed to the unpredicted event, and con-

Table 4. Crossover cavern: initial instrumentation results

	Calculated values	Measured values
Roof settlement: mm	17–32	8–63
Crown heading horizontal convergence: mm	18–23	−9 – +6
Shotcrete stress: MN/m²		
Crown	4·9–6·8	1·7–12·2
Sidewall	8·4–13·2	0·4–4·2

firmed that the excavation was still stable under such conditions.

66. Progress rates improved during cavern construction, delayed only by the unpredicted event which prompted the installation of drainage holes 6 m long in the cavern crown along its length. Despite this improvement, the first running tunnel TBM entered the cavern before construction of the central invert, except for 20 m lengths at each end which incorporated the cross-passage connections passing to the service tunnel to the north (Fig. 16).

67. All mass concrete and shotcrete was supplied by rail from the surface 7 km away, but secondary lining required the installation of a batching plant within the cavern. The secondary lining was poured as one complete arch of a 5·5 m bay length with each bay requiring 165 m³ of 40 MN/m² concrete.

Fig. 16. Entry of RTN TBM into crossover cavern

Fig. 17. Completed secondary in situ concrete lining in crossover cavern prior to installation of internal structures

68. The as-built wriggle alignment survey of the cavern secondary lining together with the four legs of the running tunnels required an adjustment of only 35 mm to the theoretical centreline to which the remaining segregation walls, crossover door, catenary, track and diamond crossing were constructed (Fig. 17).

Conclusion

69. The design and construction of the major structures addressed in this Paper were subject to change resulting from the evolving nature of the functional requirements. Changes to the construction method to optimize the programme were achieved efficiently by close co-operation between all parties, and by utilizing the inherent advantages of the NATM system.

Acknowledgements

70. The Authors are grateful to Eurotunnel for their permission to publish this Paper, and to their colleagues in TML and their consultant Mott MacDonald and sub-consultant ILF for their contribution to the design and construction of the works described.

References
1. BRITISH STANDARDS INSTITUTION. *BS 8110: Structural use of concrete.* BSI, London, 1985.
2. *DIN 1045: Beton und Stahlbeton—Bemessung und Ausführung* (Reinforced concrete structures—design and construction). Beuth Verlag GMBH, Berlin, 1988.
3. HOEK E. and BROWN E. T. *Underground excavations in rock.* Institution of Mining and Metallurgy, London, 1980.
4. ZIENKIEWICZ O. C. *The finite element method.* McGraw Hill, London, 1963, 3rd edn.
5. PÖTTLER R. Time-dependent rock-shotcrete interaction: a numerical shortcut. *Comput. geotech.,* 1990, **9**, 149–169.
6. POLLARD C. *et al.* Construction planning and logistics. *Proc. Instn Civ. Engrs, Civ. Engng, Channel Tunnel: Part 1—Tunnels,* 1992, 103–126.
7. PENNY C. *et al.* Castle Hill NATM tunnels: design and construction. *Tunnelling '91.* Elsevier Applied Science, London, 1991, 285–297.
8. FUGEMAN, I. C. D. *et al.* Development of the design and construction methods for the UK undersea crossover. *Tunnelling '91.* Elsevier Applied Science, London, 1991, 427–439.
9. BIRCH G. P. *et al.* Geotechnical aspects of the design and construction of the UK crossover. *Tunnelling '91.* Elsevier Applied Science, London, 1991, 143–159.
10. HAWLEY J. and PÖTTLER R. The Channel Tunnel: numerical models used for design of the UK undersea crossover. *Tunnelling '91.* Elsevier Applied Science, London, 1991, 441–449.
11. ITASCA CONSULTING GROUP. FLAC version 2.20: verification, example and benchmark problems. *US Nuclear Regulatory Commission Report No. NUREG/CR-5430,* 1989.

Construction planning and logistics

C. Pollard, BSc(Eng), FICE, T. J. Green, BEng, FICE,
and R. G. Conway, BSc(Eng), MICE

Proc. Instn Civ.
Engrs, Civ. Engng,
Channel Tunnel,
Part 1: Tunnels
1992, 103–126

Paper 9934

■ In general terms the UK tunnel construction can be considered as a £1000 million design-and-build project to be completed in 60 months. The fast-track nature of the operations required a system for planning and logistical control which covered the needs of overall strategic planning and detailed day-to-day planning for long linear work sites with limited access. This Paper illustrates the systems used in the context of the early construction start-up, the construction of a major seawall and the development of a working site using tunnel spoil starting from an extremely limited area on the coast near Dover. It describes the operation from this site of the third largest rail network in the UK, supporting six high-speed tunnel drives and associated tunnels, five of them simultaneously with crossover and pump station construction, and later the permanent fixed equipment installation.

Overall project planning

The management of the construction of a multi-discipline fast-track design-and-build project of the scale of the Fixed Link requires a planning system which provides essential overall strategic planning, detailed planning and reporting to management to ensure successful completion. Even with current computer capacity, there is a danger of creating programmes of such complexity that they fail to meet the needs of flexibility and adaptability. An overall hierarchical planning and control system was created to keep programmes to manageable sizes (see Fig. 1). The overall project programme (level 1) covered work in both the UK and France.

2. Activities were broad-based, usually covering the whole of a major subsection such as the crossover or a single tunnel drive. For general management purposes the project was divided into a number of sub-projects. In the UK these were

(a) tunnels
(b) terminals
(c) Isle of Grain, precast works
(d) transportation system and engineering (UK and France)
(e) rolling stock.

3. The tunnels sub-project was sub-divided into construction and engineering. Each sub-project produced a detailed programme (level 2). These programmes were co-ordinated with the other sub-projects through a central planning team which ensured that they complied with the project-wide strategic plan.

4. The strategic basis for the level 1 programme was a time chainage programme (see Fig. 2). This was a simple programme co-ordinating UK and French tunnelling activities. From this programme the UK tunnel construction team developed level 2 detailed programmes. These focused attention on the priority of the establishment and development of the working site at Shakespeare Cliff.

General concepts and constraints for Shakespeare Cliff site development

5. The Lower Shakespeare Cliff platform was identified in the Submission to Government as the location of the main working site for tunnelling operations. The site had been partly developed for the 1974 scheme. An access tunnel had been driven from the cliff top to the platform below, and from this platform an inclined access adit had been driven, together with a short length of service tunnel. A British Rail siding facility had also been established. However, the area was insufficient for the 1985 requirements (Fig. 3).

6. The 1985 proposal relied on the use of tunnel spoil to create the larger working area as tunnelling progressed. The minimum essential working platform required would absorb about 3×10^6 m³ of the excavated solid chalk marl. The level 2 programmes indicated a need to dispose of 4.75×10^6 m³ of solid chalk marl. Although the need to create the essential working platform was accepted by all parties, the deposition of the surplus spoil was the subject of much study and debate.[1] Some 70 alternative sites were investigated and 10 other major interested parties were consulted.

7. The debate was concluded in April 1987 and the Channel Tunnel Act (CTA) embodied the permission to place the surplus spoil at the Shakespeare Cliff site.

8. Although the CTA gave the necessary consents for the project as a whole, the details were still subject to local authority approval. For the Shakespeare site this has required a total of 121 submissions phased over five years and the involvement of 40 consultee bodies. Although this involved a considerable effort and imposed restraints on the programme of construction operations, it has ensured that the environmental aspects were fully considered.

9. Another aspect of great concern to the local authorities was the delivery of materials

C. Pollard,
Engineering
Manager —
Construction,
UK Tunnel
Construction
Transmanche-Link
JV

T. J. Green,
General
Manager —
Railways and Pithead,
UK Tunnel
Construction
Transmanche-Link
JV

R. G. Conway,
Section
Manager —
Seawall and
Infrastructure,
UK Tunnel
Construction
Transmanche-Link
JV

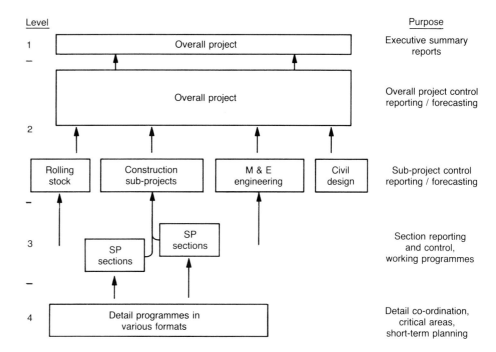

Level		Purpose

Level 1 — Overall project — Executive summary reports

Level 2 — Overall project — Overall project control reporting / forecasting

Sub-project control reporting / forecasting: Rolling stock · Construction sub-projects · M & E engineering · Civil design

Level 3 — SP sections · SP sections — Section reporting and control, working programmes

Level 4 — Detail programmes in various formats — Detail co-ordination, critical areas, short-term planning

Fig. 1. Heirarchical planning structure

to the site. The quantities are indicated in Table 1. Wherever practicable, delivery was to be by rail, and this was favoured by all concerned; rail-delivered materials were precast concrete (PCC) tunnel segments, sheet piles and aggregates. Rail delivery of other bulk materials such as cement was not possible owing to restrictions in site layouts.

10. The PCC tunnel linings varied between land and marine drives and within these drives to suit the various ground loading characteristics. In addition, there were large numbers of cast iron lining types required for both main drives and the ancillary structures. Specific programmes were developed to enable development of workable site layouts to cater for the changing lining requirements as tunnelling progressed and the site expanded.

Table 1. General statistics for tunnel construction materials handled at Shakespeare Cliff lower site

Item	Delivery mode	Pieces	Tonnes
Tunnel linings			
PCC segments (19 types)	Rail	450 000	1 646 500
SGI rings (33 types)	Road	123 000	37 000
Other materials			
Concrete aggregates	Rail	N/A	900 000
PCC service tunnel planks	Road and rail	21 000	63 000
PCC running tunnel sleeper blocks	Road and rail	142 000	47 000
Rail (180 km track)	Road	52 000	13 500
Sheet piling	Rail	17 600	24 500
Various pipes (225 km)	Road	37 500	6300
Totals		843 100	2 737 800

11. With the very limited site working area and the need to develop rapidly the management infrastructure, main offices and welfare facilities were to be established on the Upper Shakespeare Cliff site with man access to the tunnels via a shaft. Five high-speed tunnel drives plus other underground construction were to be serviced simultaneously by the lower site facilities.

12. Spoil would be removed by a conveyor installed in the existing adit (A1) and a second adit (A2) would be required to accommodate five 900 mm gauge rack tracks, one for each tunnel drive (Figs 11, 13 and 14).

13. The tunnels required power, water, ventilation and drainage services. Power was to be provided from two separate 11 kV feeds from the National Grid. These provided a total power supply of 23 MVA. In addition, 4 MW of standby generator capacity was provided; this was designed to run the tunnel lighting and pumping in the event of a mains failure.

14. Water supplies would be required to cool the tunnel boring machines (TBMs), and also for concrete, grouting and general welfare. A total capacity of over 30 l/s was required; however, the droughts of 1989 and 1990 made it impossible for the local water supplier to meet this requirement. Two desalination plants with capacities of 10 and 20 l/s were therefore installed during the construction period.

15. Ventilation requirements were complex and would vary as the job progressed. Fan stations would be established at the top of both adits and the shaft for the early tunnel drives. Eventually fan stations would also be established at the terminal portals and in mid-Channel once the tunnels had holed through.

16. Drainage facilities needed to cater for the removal of TBM cooling water, seepage flows from the rings, plus an allowance of 275 l/s for possible sudden inundation. The required temporary pumping capacity was 745 l/s. Water would need to discharge into the spoil lagoons or settling ponds before the sea.

Development of Upper Site

17. In September 1987 the only access to the site was from the west via the A20 road and the 2 km long Old Military Road. Access from the east via Dover and the Aycliffe housing estate was totally prohibited, except to emergency vehicles (see Fig. 4).

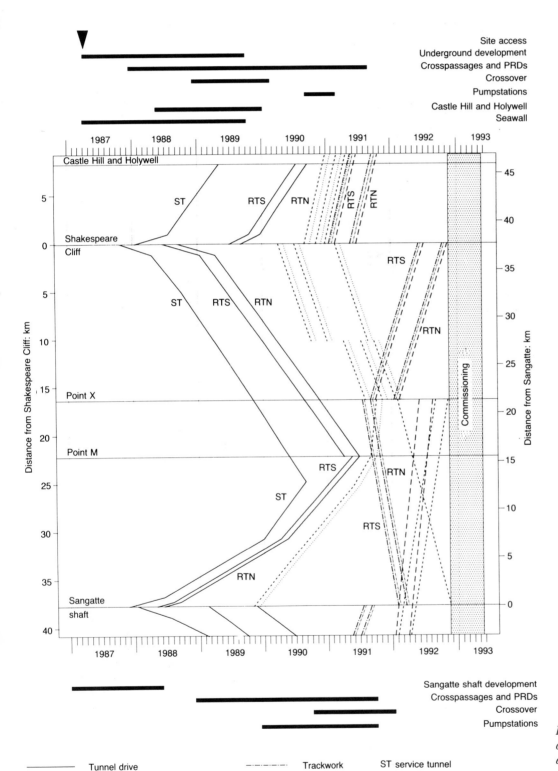

Fig. 2. Level 1 time chainage programme, as at November 1986, with significant barline activities

*Fig. 3. Lower
Shakespeare Site,
June 1987*

18.   Once planning permission had been obtained in October 1987, work immediately started on a new highway standard road from the A20, but until its completion in March 1988 access was severely limited, with vehicles travelling in convoys and heavier vehicles needing to be towed up the steep inclines.

19.   From the Upper Site access, to the Lower Site was via the existing access tunnel. This remained the only road access to the Lower Site. Personnel access to the underground workings was via a 10·0 m diameter shaft, 106 m deep and adjacent to the main changing rooms. The shaft contained three rack-and-pinion passenger hoists. These gave an available capacity of 1000 men/hour up and down the shaft.

20.   The Upper Site facilities had to cater for 900 staff and over 5000 operatives. From its humble beginnings of Portakabins partly blown away in the October 1987 hurricane, the site was developed to contain purpose-made office blocks, a visitor centre, a control and communications centre, changing facilities, a medical centre, a canteen, a bank, a time office, a weighbridge and facilities for the police and fire brigade. Customs and Excise staff were also accommodated when the site became an official port of entry into the UK in 1990. These facilities were maintained around the clock on each day of the year.

*Fig. 4. Upper Site
facilities*

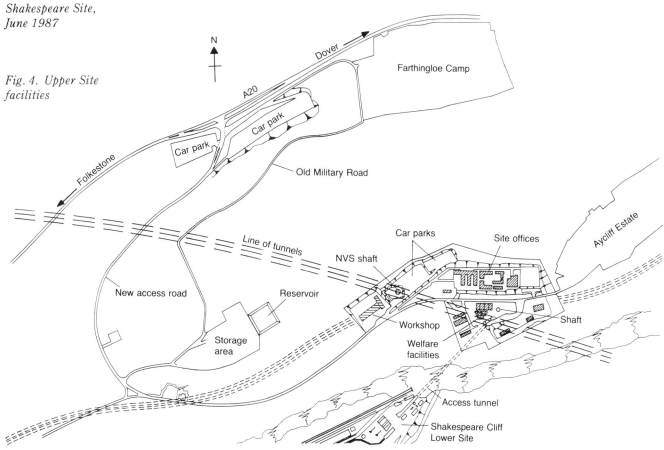

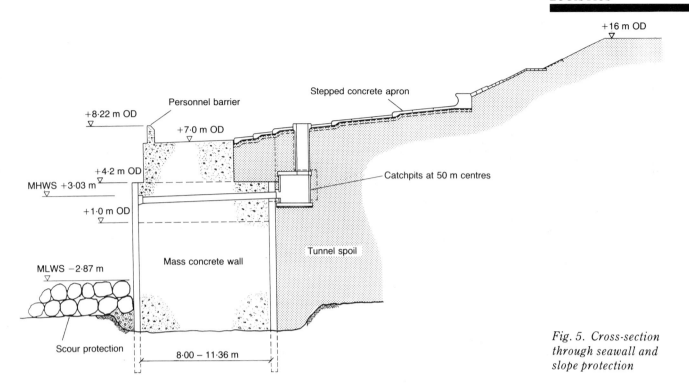

Fig. 5. Cross-section through seawall and slope protection

21. To cater for car parking, areas with a total capacity for nearly 1000 cars were established adjacent to the A20. A timetabled fleet of buses transported workers from here and the nearby construction village to the Upper Site facilities.

Lower Site development

22. In parallel with the Upper Site development the programme required an immediate and simultaneous start on the construction of the shaft, the new adit (A2), the conveyor tunnels at the bottom of A1 and preparations for erecting the first TBM. As there was insufficient space available for spoil disposal, this meant that seawall construction also had to start immediately.

Seawall construction

23. Under the terms of the Channel Tunnel Act, measures had to be adopted to ensure that none of the chalk spoil placed within the Shakespeare Cliff Lower Site area leached into the sea.

24. It was also necessary to design and construct a seawall which would allow rapid construction to meet the tunnelling programme and withstand storm damage at any stage during its construction.

25. Since the seawall was critical to the overall success of the project, model-testing, marine, feasibility, environmental and risk-analysis studies were instigated at the outset. If the seawall construction had been delayed by storm damage, with nowhere to put tunnel spoil, tunnelling would have inevitably

stopped. This dictated a conservative approach, using tried-and-tested methods, experienced labour and staff, readily available materials, suitable marine plant, and finally (and perhaps most importantly) substantial temporary works.

26. The final adopted solution was a 1700 m long seawall of mass concrete, contained by a double row of sheet piles, and on average 200 m from the shore (Fig. 5). A series of gravel-filled crosswalls formed lagoons to contain and settle chalk fines (see Fig. 6). Five lagoons were constructed and their positions along the seawall were dictated by tunnel progress to ensure that no chalk spoil was deposited into a lagoon until it was closed. Separate detailed programmes were produced to provide anticipated spoil production figures and minimize the number of crosswalls.

27. Although the contract programme showed the start of UK service tunnel boring operations in December 1987, other necessary site developments generated spoil much earlier. It was not possible to obtain the necessary planning permissions for the permanent concrete-filled wall within the short timescale; it was therefore decided to construct a temporary works reclamation (TWR) (see Figs 7, 8 and 11) which could be removed on completion.

28. Site access was granted at the end of August 1987, some 4 months after commencement of the contract. Despite the October hurricane and subsequent storms, an initial single sheet piled lagoon was completed in mid-November 1987 (Fig. 7) and the total TWR in February 1988.

107

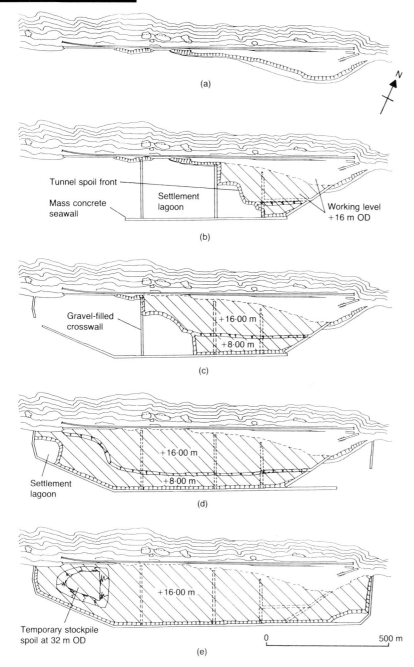

Tunnel spoil front

Mass concrete seawall

Settlement lagoon

Working level +16 m OD

(b)

Gravel-filled crosswall

+16·00 m

+8·00 m

(c)

Settlement lagoon

+16·00 m

+8·00 m

(d)

Temporary stockpile spoil at 32 m OD

+16·00 m

0   500 m

(e)

Fig. 6. Construction development of Shakespeare Cliff land reclamation: (a) August 1987; (b) April 1989; (c) January 1990; (d) October 1990; (e) June 1991

29. The seawall consists of a double row of Larssen 6 steel sheet piles, varying in distance between the piles from 8·00 to 11·36 m, depending on the level of the hard chalk below the seabed. With sheet pile diaphragms at intervals between them, a series of 10·08 m cofferdams was created.

30. The construction sequence was as follows: to dredge a trench along the line of the wall to remove the 1·0–1·5 m of superficial deposits; to drive the sheet piles to form the cofferdam; to airlift the debris using divers; and to place the first lift of concrete underwater from approximately −5·0 to +1·0 m OD.

31. A second lift of mass concrete was

poured in the dry at low water up to +4·20 m OD after removal of the piling frames, which also acted as supports during storm conditions. Above this level the wall width reduced to a constant 6·60 m, and using a travelling shutter the mass concrete was poured to +7·0 m OD.

32. The innovative design of the 5·04 m long piling frames played a major part in the success of the construction (see Fig. 9). Their basic functions were to resist loading due to hydrodynamic forces imposed by wave action, to resist loading due to pressures imposed by wet concrete, and to act as a guide for driving the sheet piles forming the wall faces.

33. Concrete for the seawall was provided by two Stetter batchers, each with an output of 80 m³/h. In order to meet the requirements of non-reactive mixes in the tunnels and to minimize storage requirements, all concrete was produced using Glensander crushed granite, granite sand, pulverized fuel ash (PFA) and cement.

34. Glensander aggregates were delivered by sea from Scotland to a facility adjacent to the Isle of Grain precast segment factory. From here they were delivered by rail to Shakespeare Cliff.

35. In all 180 000 m³ of concrete were pumped into the seawall over distances up to 500 m. Initially a concrete pump was situated under the batching plant, and as work progressed another pump was introduced as a stage pump. Subsequently, truck mixers transported concrete along the top of the new seawall to pump stations which moved as the work progressed.

36. Generally, two S70 Flexifloat jack-up platforms with Manitowoc 4000 crawler cranes were used to construct the seawall. One concentrated solely on driving piles, commencing in April 1988, and on average it progressed at the rate of 9·0 m double sheet piled wall per week in winter and 13·5 m per week in summer. The other barge was required to airlift at high tide, move the concrete placing boom, place tie rods and extract piling frames (Fig. 10).

37. Crosswalls were constructed using a third jack-up platform and Manitowoc 4000 crane, in conjunction with a 800 SC crawler crane working from the land on the as-dredged ballast placed between the sheet pile walls.

38. Safety was paramount. A rigid-hull inflatable safety boat operated on a 24 hour basis and was manned by rescue-trained divers, who safely recovered 6 personnel over a 4 year period.

39. A 1000 tonne buffer stock of piles was maintained on site. Piles were loaded onto a flat-top pontoon as required and towed to the jack-up platforms.

40. Piles on the rear face of the seawall and the crosswalls were extracted and reused. On average 1·5 m of pile was buckled when

Fig. 7. Lower
Shakespeare
platform, November
1987, showing first
lagoon, TWR
construction
commenced and
stockpiled 1974
segments

Fig. 8. Lower
Shakespeare
development, June
1988, filling of TWR
nearing completion
and permanent wall
started

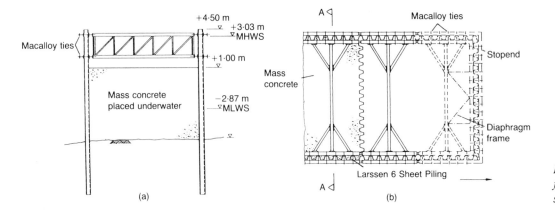

Fig. 9. Piling frame
for seawall: (a)
section A-A; (b) plan

Fig. 10. Seawall construction in October 1989 during storm

extracted. In total 32 500 tonnes of piles were driven using a Delmag D46 or IHC S70 hydraulic hammer, of which 8000 tonnes were extracted using a modified ICE 1412 vibrating hammer and 250 tonne clamp.

Land development and pithead area

41. The total area of the Lower Site to be reclaimed within the seawall area amounted to 34 hectares, of which 5 hectares were for the permanent buildings at the east end of the site.

42. In early 1988, when the tunnels were in difficulties owing to ground conditions, a critical situation developed. More back-up facilities were required to improve production, but without the tunnel spoil land could not be reclaimed on which to construct the facilities. To ease the problem small temporary workshops were established until space became available for the main ones, and more facilities were established on an extension to the Upper Site. Eventually it was decided that land development should proceed at +16 m (the main working level) and at +8 m. This allowed a much more rapid development of valuable working and storage space. Liaison between all departments at this stage was critical, and development drawings showing the Lower Site at monthly intervals were continuously updated by the planning department to co-ordinate site requirements.

Fig. 11. Site layout, early 1988

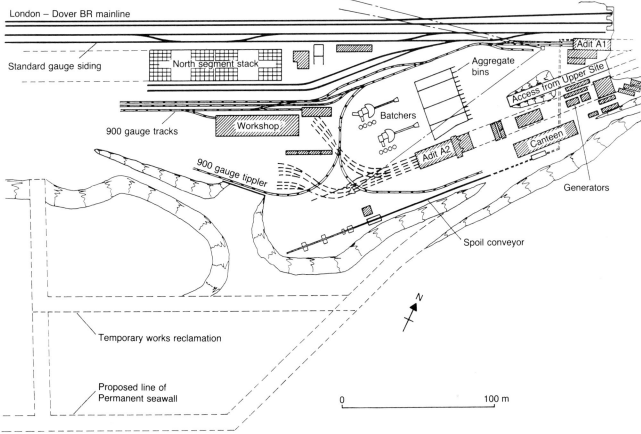

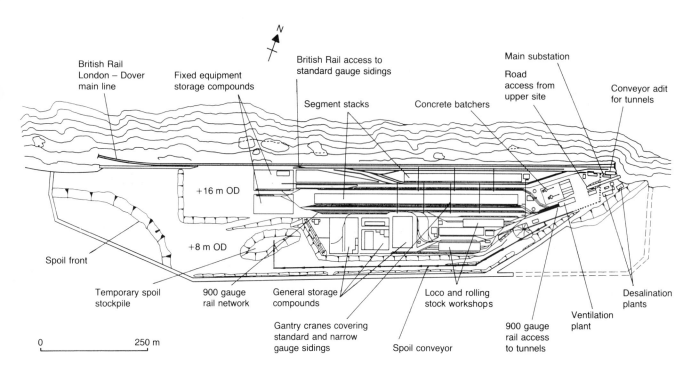

British Rail London – Dover main line

Fixed equipment storage compounds

British Rail access to standard gauge sidings

Segment stacks

Concrete batchers

Main substation

Road access from upper site

Conveyor adit for tunnels

+16 m OD

+8 m OD

Spoil front

Temporary spoil stockpile

900 gauge rail network

General storage compounds

Gantry cranes covering standard and narrow gauge sidings

Spoil conveyor

Loco and rolling stock workshops

900 gauge rail access to tunnels

Ventilation plant

Desalination plants

0   250 m

Fig. 12. Shakespeare Cliff Lower Site layout, June 1990

Fig. 13. Lower Site facilities, June 1990

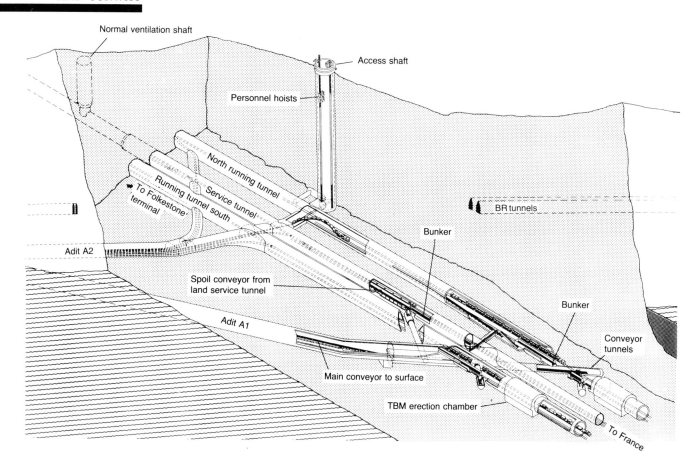

Fig. 14. Shakespeare Cliff underground marshalling area

Fig. 15. 14 m³ Mulhauser muck skip discharging in marshalling area bunker

43. Early priorities were to extend the British Rail sidings, and to provide workshops, a segment storage area, batching plants and a sheet pile laydown area. The north segment stack initially provided materials for the 900 gauge rail tracks into adit A1. At a later stage north and south stacks would serve 5 tracks into adit A2 (Figs 11 and 12).

44. Each of the five rack tracks in adit A2 was serviced from a designated stacking area. To accommodate programme needs and the requirement for a minimum four weeks of stockholding, two main segment stacks were required. Each was serviced by its own British Rail siding involving 3 km of standard gauge track.

45. Each siding and each tunnel drive was allocated a 20 t high-speed gantry crane, giving a total of 7 cranes. These were required to have a loading cycle of under 3 minutes per segment, including turning the segment through 90° for placing on the 900 gauge wagons.

46. Rapid unloading of the 2000 t BR trains to meet the required turnaround times was achieved by using multiple segment lifts and having all keys palletized. Segments were stored and transported on edge in the BR wagons. BR train loading was carefully planned to achieve a high load factor, and to suit the precast factory manufacturing and stacking arrangements and the Shakespeare Cliff stacking and transhipment arrangements. As a general policy, the planning was in 'rings' of segments.

47. The final reclaimed area taken up by the Pithead area alone was 11 hectares, which involved the laying of 56 000 tonnes of scalpings and 12 km of 900 gauge railway track (see Figs 12 and 13). In June 1990 reclamation was sufficiently advanced to allow development of a further 2·5 hectare area, to the west of the Pithead, for fixed equipment installation requirements. Substantial settlement of the reclamation area was predicted and up to 500 mm

was observed. Settlement under the 20 t gantry crane rails was of particular concern. The chosen rail support was steel sleepers on ballast to allow the rails to be re-levelled as the settlement caused by the 9·0 m high PCC segment stacks occurred.

48.   The production of tunnel linings at the Isle of Grain, and various foundries and the production of PCC track support units, were not easily changed to suit space requirements at the Lower Shakespeare Site. It was therefore found necessary to develop off-site storage facilities. At Sevington, near Ashford, a rail-head had been established for terminal construction materials. This was extended by 3·5 hectares to give a facility to provide a buffer store for materials. A second off-site storage was also established for use as a buffer stock for PCC tunnel segments. At peak it contained 2000 rings.

Spoil handling

49.   Below the marshalling area (Fig. 14) 3 and 4 m dia. tunnels at the bottom of adit A1 contained feeder conveyors to receive spoil from the tunnel bunkers and meter it on to the 2400 t/h main inclined conveyor. A muck train of twelve 14 m³ skips could be discharged into a bunker in 5 minutes (Fig. 15). At the top of the adit, the conveyor fed an underground conveyor to the seaward edge of the site, from where other conveyors transported it to the discharge point.

50.   The critical nature of the varying tunnel output required that resources were available to meet predicted peaks. Initially 15 tonne dump trucks were loaded at the east end of the site by rubber-tyred loaders from the stockpile at the end of a fixed conveyor (Fig. 16).

51.   The dump trucks initially end-tipped spoil into the lagoon to displace the sea water through the crosswall ballast which acted as a filter. Between +11·0 and +16·0 m OD, 90% compaction was obtained by spreading and compacting the spoil in 300 mm layers and 16 passes of a Cat D6 and 6 tonne sheepsfoot vibrating roller. The net bulking factor from solid in situ to compacted fill was continuously monitored, and it averaged 1·24.

52.   In 1988 in order to develop land area for both facilities and conveyor installation with the limited amount of spoil being produced, the site was developed at both +8 and +16 m OD. In April 1990 tunnel production increased to the extent that at times spoil generation exceeded the conveyor design capacity of 2400 t/h and the eight dump truck and two Cat 966 loading shovels were only just able to cope. The decision to develop the site on two levels had by now generated sufficient area to allow the conveyor system to be extended 600 m westward and for a radial speader conveyor to be added to the head end (Fig. 17). This not

only reduced the haul distance for the dump trucks, but it also increased the stockpile capacity from 5000 to 25 000 tonnes. To save operating costs, the existing eight smaller dump trucks were replaced by 3 No. D400 25 tonne dump trucks and 1 No. Cat 980 wheeled loader. Although muck shifting normally stops during the winter, output from the tunnels continued unabated 24 hours a day for nearly 4 years.

53.   In December 1990 the filling of the fifth and final lagoon at the eastern end of the site commenced; this coincided with the rundown in

Fig. 16. Fixed spoil discharge and transfer conveyor to radial spreader

Fig. 17. Radial spreader conveyor

Fig. 18. Aerial view from the west showing site and spoil stockpile March 1991

Fig. 19. Preliminary proposals for landscape works at Shakespeare Cliff Lower Site incorporating amenity and nature conservation features

tunnel production. By April 1991 the filling of this lagoon was complete and tunnel production was sufficiently reduced to dispose of the 25 t dump trucks and radial spreader conveyor, and move to single-shift working. At this time over 28 hectares had been created, together with a 500 000 m³ stockpile for future landscaping (Fig. 18).

54. It is envisaged that in early 1993 demobilization of the Lower Site will have advanced to the stage when muckshifting of the stockpiled material may commence in order to create the final landscaped amenity and conservation area, ready for seeding in September 1993 (see Fig. 19).

Operation of pithead facilities

55. During the construction activities a separate organization concentrated on servicing tunnelling activities. These people were dedicated to and accountable for safe and efficient loading and unloading of trains, speedy turnaround of rolling stock and delivery of the goods underground. They had a vital role of taking delivery of the goods ordered by others, keeping meticulous stock control checks and breaking down bulk materials into mixed train loads for use at the working faces.

56. On a tonnage basis the majority of materials used on the project was delivered to site by rail. The major items were the tunnel linings; these involved 13 types of PCC rings and 33 different SGI rings. To maximize storage capacity the segments were stored on edge up to 6 segments high (9 m).

57. Almost 450 000 segments totalling over 1 646 500 t were delivered, and each was handled twice. In all, some 35 000 different types of materials were delivered, in sequence, to prearranged timescales to the 50 or so work sites active at any one time underground.

58. Both the unloading and loading of trains was performed by teams of 4 men, two slinger banksmen on the stacks and two on the trains. For despatch into the tunnels, the segments were carried horizontally, intrados upwards, on 6 m long flat cars. Each train also carried the ancillary materials such as bolts, rails, pipes and packings.

59. At peak times 50 concrete remixer cars, some of 7 m³ and some of 9 m³ capacity, were used to take pre-batched wet concrete underground by rail, often to over 20 km from the batching point (see Fig. 20).

60. In the latter stages of the project the Pithead role was reversed. It became the staging post for the collection, sorting and disposal of the many temporary materials and equipment which had been used underground. During this workphase the fixed equipment installation sub-contractors who had their own

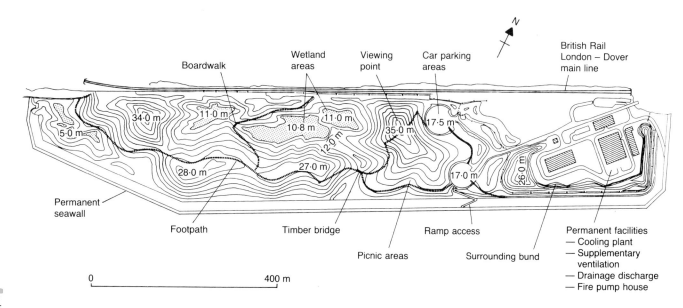

yards were also serviced by the construction
railway via the Pithead.

Operation of construction railway

Development of method of management

61.   It was realized at an early stage that the
construction railway and the whole logistic
cycle for delivery of men and materials and the
removal of spoil would be crucial to the project.

62.   It was necessary to serve 5 TBMs operating simultaneously, and also to construct
over 150 pairs of cross-passages and over 100
piston relief ducts, two major pumping stations
and the UK crossover. Many miles of temporary
services had to be placed and later removed,
and a service needed to be provided to the sub-
contractors responsible for the installation of
the permanent mechanical and electrical services. At peak production some 1000 men had
to be taken to and from work in the tunnel, at
each of the three major shift changes each day.

63.   Each tunnel contained a double-track
37 kg/m 900 mm gauge construction railway
with pairs of facing and trailing crossovers at
375 m intervals. The pit bottom area where
these 12 tracks were connected to the 5 track
rack railway in access adit A2 was complex
(see Fig. 21).

64.   The combination of the complexity of
the layout, the size of the operation (over
150 000 train km per month at peak and some
$2\frac{1}{2}$ million train km throughout the project), and
the 160 locomotives and the 990 units of rolling
stock (see Tables 2 and 3 and Figs 22–26)
required that the railway be managed as an
entity. An organization called Underground
Operations (UGO) was formed which was independent of the tunnel construction teams and
provided a service to them. The tunnel construction teams were the 'customers' of UGO.
This concept is unusual in tunnelling works,
and it took time to evolve and to establish an
effective working method. Prioritization of conflicting logistic demands became the customer's
business, and UGO ran the railway within the

frequently changing priority framework established by its customers.

65.   Overall priorities were established by
the construction planning team. From their
overall level 2 strategic programmes, they produced detailed time chainage programmes for
the tunnel drives and networked level 3 bar
charts for each tunnel and major structures
such as the marshalling area, pump stations
and the crossover. In the early stages of design
development, these only covered a 9 month
period, but as design progressed they were
extended to cover the full activity. From these
detailed level 3 programmes the level 2 logic
was further refined.

66.   The detailed time chainage programmes
were initially developed for prioritization in the
service tunnel (see Fig. 27). Cross-passages
were required to house sub-stations and temporary pump stations. There was a limit to how
close cross-passage construction could get to
the TBM, and there were limits to how far the
TBM could advance from the last installed substation or pump station. The installation of
trackwork was also controlled by these activ-

*Fig. 20. Lower
Site—south segment
stack with two
remixer cars and
loaded segment train
behind*

*Fig. 21. Shakespeare
Cliff underground
marshalling area*

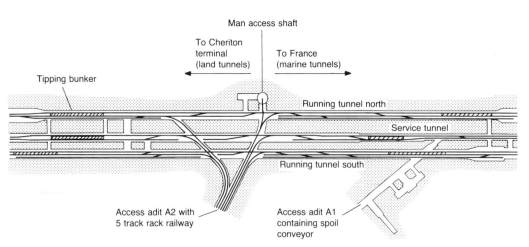

Fig. 22. Mark II Tandem Hunslet Battery charging from the 550 V DC overhead line

Fig. 23. Schoma CFL 180 diesel locomotive hauling half-rake of muck skips

Fig. 24. Schoma 150 diesel locomotive on arrival at site

Table 2. Locomotive fleet

Type	Quantity	Comments
Electric locos (construction railway 900 mm)		
Mark I 175 HP Hunslett battery/catenary adhesion	20	10 of these converted to diesels in 1991
Mark II 200 HP Hunslett tandem battery/catenary	40	Connected as 20 tandem pairs
350 HP Hunslett battery/ catenary rack and pinion	18	
Diesel locos (construction railway)		
NCB style rack/pinion	2	
185 HP Schoma CFL 180	33	
200 HP Schoma tandem rescue CFL 200	4	2 tandem pairs used also for normal work
150 HP Schoma CFL 150	18	
150 HP RFS shunters	7	Used mostly on surface at pithead
112 HP Simplex shunters	2	Duties confined to workshop area
200 HP Hunslett tandems	10	5 tandem pairs formerly Mark I electrics
		8 of these converted a second time to standard gauge locomotives
Manriders (construction railway)		
80 HP Schoma D60 diesel manriders (2 and 3 car sets)	16	Total self-propelled manriding capacity 784 seats
Other prime movers (construction railway)		
Self-propelled flat cars move 4 m by ram stroke	6	In-house prototype; production by Schoma
Self-propelled flat cars move 100 m by winch	9	
Standard gauge locomotives		
420 HP Brush TM68	5	Used from late 1991 to serve customers not
200 HP Schoma conversions of Hunslett diesels	8	contracted to supply their own locomotives

Table 3. Rolling stock fleet

Type	Quantity	Comments
Muck skips (construction railway 900 mm)		
Mulhauser side tipping 14 m^3 skips	333	70 converted to flat beds after muck
Mulhauser side tipping 11 m^3 skips	10	shifting substantially completed
Flat beds (construction railway)		
Material cars (Mulhauser)	266	Majority of fleet was air-braked, 12 of these
Material cars (Becorit)	288	flat beds were converted to standard gauge
Platform cars unbraked	57	rolling stock in late 1991
Cable drum cars	14	
Cement cars	11	
Mess/toilet cars	31	
6 m flat beds created by conversion from skips	70	
Concrete remixer cars (construction railway)		
7 m^3 Mulhauser	33	Used mostly to convey wet concrete batched
9 m^3 Mulhauser	9	at surface, taken down rack and as far as 20 km
9 m^3 Becorit	9	from pit bottom
Manriding cars (construction railway)		
Mulhauser	34	These cars were marshalled in locomotive-hauled material and muck trains
Standard gauge rolling stock		
10 m long hired flat beds	32	Used from late 1991
6 m long converted construction railway flat beds	12	

Fig. 25. Hunslet Rack Loco in adit A2

Fig. 26. Schoma Manrider

ities; these constraints could be clearly seen in the time chainage format.

67. At a more detailed level, a four-week rolling programme was prepared by the planning department. This covered all tunnel activities and was of particular importance during the start-up period when 6 TBMs had to be erected with various heavy lifts, coupled with changing access requirements as they were launched. Weekly co-ordination meetings were held and the programme was issued on Wednesdays; at peak, it was issued to over 90 people and contained over 250 activities.

68. Despite considerable detailed planning, decisions on the quantity of logistical support often appeared, prima facie, to have been underestimated. Lead times were such that the short-term provision of extra resources was rarely possible. That such difficulties were overcome by better utilization of existing resources and pragmatic short-term replanning is a tribute to all involved in these activities.

69. At times, opportunities to flood a tunnel with resources were available, for example in running tunnel south when the north marine TBM was moving through the crossover. However, these situations usually demonstrated that congestion caused by the necessary layout of the railway, and the inevitable sections of single-line working needed by other activities, were the limiting restraint.

Locomotive and rolling stock fleet

70. The strategic choice of locomotives and rolling stock was based on the need to provide segments and remove spoil for construction of one ring in the running tunnel drives and two in the service tunnel drives, each using one train.

71. Environmental considerations led to the initial choice of a 550 V DC overhead supply as a principal source of locomotive power, with the support of a very small number of diesel locomotives for emergency use. Electric power was provided via twin shrouded aluminium/stainless steel conductors and an inserted head pantograph. The movement of the locos relative to the conductors and saline water ingress gave many teething problems.

72. Intensive studies, including video analysis of pantograph performance and practical on-site development eventually overcame these difficulties. However, a side effect was a reconsideration of the use of diesel locos.

73. A vital factor in the development of the use of diesel locomotives was tunnel ventilation. The service tunnel was the longest single-access tunnel ever driven. For a distance of 15 km ventilation was provided via a 1·2 m diameter duct with stage fans at 3 km intervals. Beyond this, air was fed from the running tunnels to the service tunnel via a piston relief

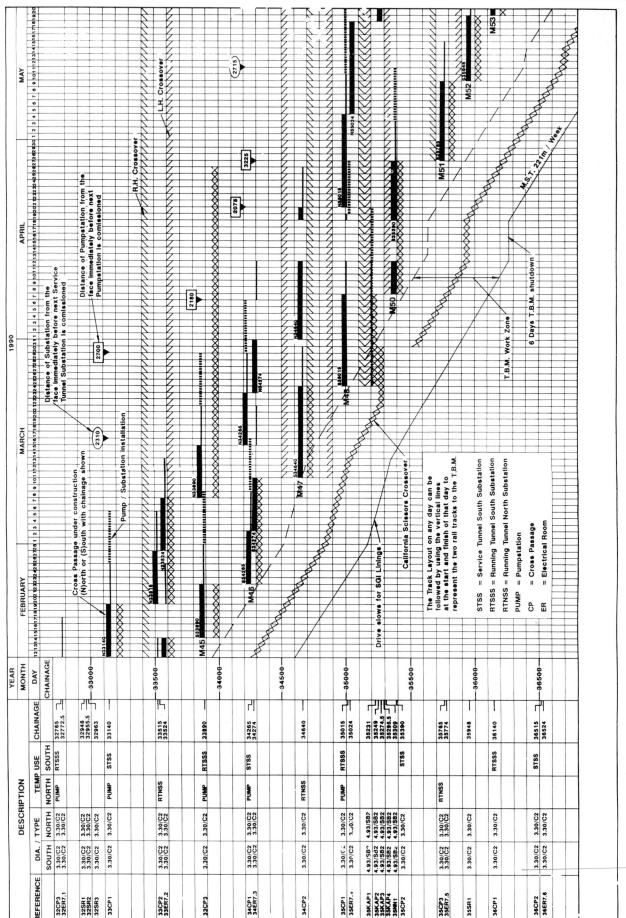

Fig. 27. Extract from 6 month service tunnel level 3 time chainage programme (tunnelling phase)

Fig. 28. Mulhauser muck skips under maintenance in rolling stock workshop

duct to maintain the dead heading to a manageable length.

74. In the running tunnels the initial ventilation design used 2·4 m dia. ducts with stage fans at the A1 portal and inbye of the crossover. This was later changed to a full-face mines-type system to the crossover, with ducted systems beyond. These developments in the ventilation design enabled the proportion of reliable, flexible, diesel locomotives to be increased. The land running tunnels and the last 10 km of drive in running tunnel north marine was achieved using diesel traction only. At this time the ventilation system in a blind-ended tunnel supported seventeen 185/200 HP diesel locomotives.

75. Twenty-four hour working, and the need for maximum availability, required in-house maintenance. This achieved locomotive and rolling stock availabilities of 90·53 and 95·56%, respectively. It also allowed TML to develop both prime movers and wagons to meet the need for incessant use with minimum breakdowns in the particularly harsh environment, and to develop specialized equipment.

76. Preventative and other maintenance of locomotives and rolling stock was planned and recorded on computer using the COMAC software system. The TML fleet was the second largest use of this system, only British Airways being larger.

Operation and control

77. Congestion of railway traffic in the tunnel complex was inevitable. It was at its worst in the marine service tunnel in early 1990, while the TBM was driving simultaneously with a major grouting operation, pump station excavation, UK crossover construction and four cross-passage construction and installation sites.

78. In the marine running tunnels congestion peaks occurred in early 1991 with the coincidence of the TBM tunnel drives averaging over 350 m/week, construction of UK crossover lining and internal structures, up to eight tunnel finishing gantries, cross-passage and PRD construction, and the start of fixed equipment installation.

79. In the marshalling area congestion was at its worst in mid-1991 when the layout was effectively single-tracked as the permanent linings were placed and fixed equipment installation was in progress. At peak times, in each of the three marine tunnels, train movements of 30 inbye and 30 outbye each eight hour shift occurred, with up to one-third of the tunnel subject to single-line working. In such traffic, incidents such as derailments, bunker and locomotive breakdowns were very significant.

80. An early decision was that an electrical signalling system on the construction railway would have been very costly and likely to fail owing to the harsh environment. Traffic was therefore controlled by one man in each tunnel, the operations board controller (OBC). He was equipped with a magnetized track layout diagram upon which he moved tokens representing the numbered locomotives. He received information from drivers and gave them instructions by radio; the frequency used was exclusive to railway movements control. In very busy areas a 'person in charge' controlled local movements to ease the burden on the tunnel OBC.

81. Efficient railway operations meant minimizing incidents, and a key contribution to tunnel productivity was the proactive management of safety within UGO. Easy-to-follow procedures for operating the railway were published. Formal in-house training of loco drivers and other railway operatives, using professional operatives rather than trained teachers, took place; electric and diesel driving cab simulators were used, each with the capacity for the instructor to introduce faults and to show videos of progress down a tunnel with staged events.

82. All incidents such as derailments and accidents to personnel were investigated to ascertain the causes and address prevention. As a measure of the success of this strategy, the number of incidents was reduced from 1·67 per 1000 train km to 0·75 over a 12 month period.

83. In line with TML's general policy each UGO operative (750 in all) received a field safety talk once a fortnight, usually on a one-to-one basis. These talks gave management the opportunity to put over safety and other messages. One-to-one safety audits were also carried out, during which supervisors encour-

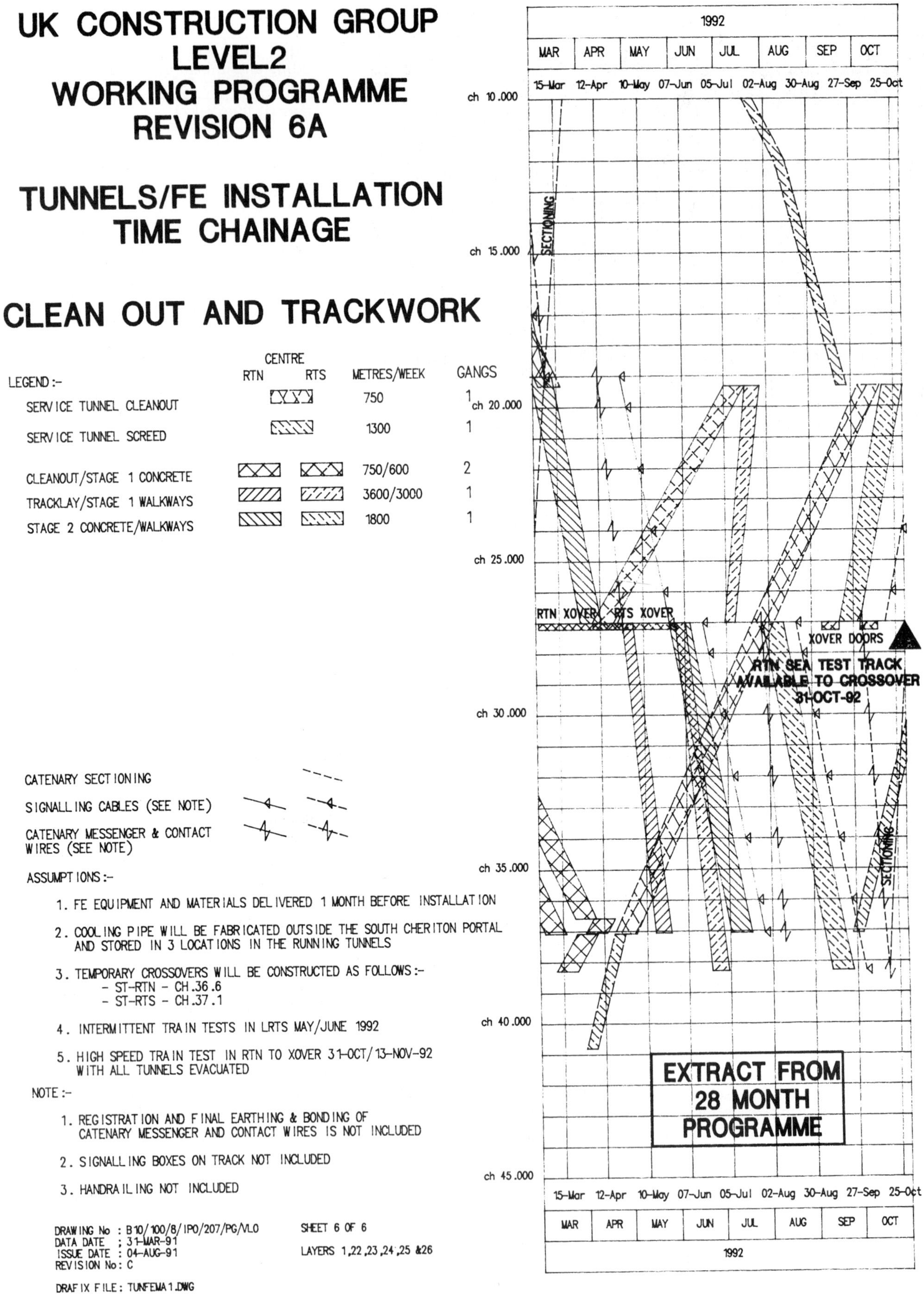

UK CONSTRUCTION GROUP
LEVEL2
WORKING PROGRAMME
REVISION 6A

TUNNELS/FE INSTALLATION
TIME CHAINAGE

CLEAN OUT AND TRACKWORK

LEGEND :-	CENTRE		METRES/WEEK	GANGS
	RTN	RTS		
SERVICE TUNNEL CLEANOUT			750	1
SERVICE TUNNEL SCREED			1300	1
CLEANOUT/STAGE 1 CONCRETE			750/600	2
TRACKLAY/STAGE 1 WALKWAYS			3600/3000	1
STAGE 2 CONCRETE/WALKWAYS			1800	1

CATENARY SECTIONING

SIGNALLING CABLES (SEE NOTE)

CATENARY MESSENGER & CONTACT
WIRES (SEE NOTE)

ASSUMPTIONS :-

1. FE EQUIPMENT AND MATERIALS DELIVERED 1 MONTH BEFORE INSTALLATION

2. COOLING PIPE WILL BE FABRICATED OUTSIDE THE SOUTH CHERITON PORTAL
AND STORED IN 3 LOCATIONS IN THE RUNNING TUNNELS

3. TEMPORARY CROSSOVERS WILL BE CONSTRUCTED AS FOLLOWS :-
   - ST-RTN - CH.36.6
   - ST-RTS - CH.37.1

4. INTERMITTENT TRAIN TESTS IN LRTS MAY/JUNE 1992

5. HIGH SPEED TRAIN TEST IN RTN TO XOVER 31-OCT/13-NOV-92
WITH ALL TUNNELS EVACUATED

NOTE :-

1. REGISTRATION AND FINAL EARTHING & BONDING OF
CATENARY MESSENGER AND CONTACT WIRES IS NOT INCLUDED

2. SIGNALLING BOXES ON TRACK NOT INCLUDED

3. HANDRAILING NOT INCLUDED

DRAWING No : B10/100/8/IPO/207/PG/VLO SHEET 6 OF 6
DATA DATE : 31-MAR-91
ISSUE DATE : 04-AUG-91 LAYERS 1,22,23,24,25 &26
REVISION No : C

DRAFIX FILE : TUNFEMA1.DWG

EXTRACT FROM
28 MONTH
PROGRAMME

Fig. 29. Extract from level 2 time chainage programme: cleanout and trackwork

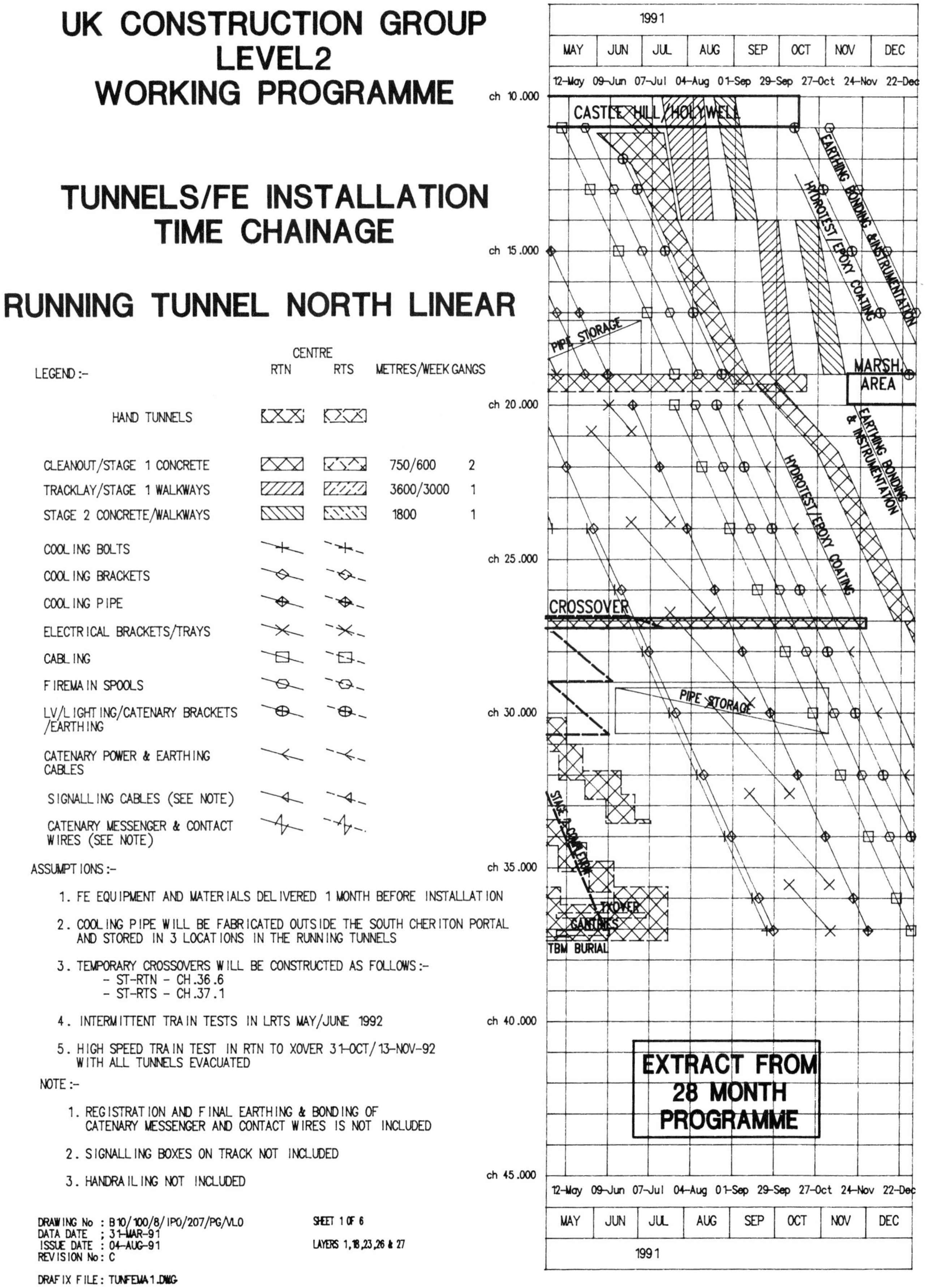

UK CONSTRUCTION GROUP
LEVEL2
WORKING PROGRAMME

TUNNELS/FE INSTALLATION
TIME CHAINAGE

RUNNING TUNNEL NORTH LINEAR

CENTRE

LEGEND :-

	RTN	RTS	METRES/WEEK	GANGS
HAND TUNNELS				
CLEANOUT/STAGE 1 CONCRETE			750/600	2
TRACKLAY/STAGE 1 WALKWAYS			3600/3000	1
STAGE 2 CONCRETE/WALKWAYS			1800	1
COOLING BOLTS				
COOLING BRACKETS				
COOLING PIPE				
ELECTRICAL BRACKETS/TRAYS				
CABLING				
FIREMAIN SPOOLS				
LV/LIGHTING/CATENARY BRACKETS /EARTHING				
CATENARY POWER & EARTHING CABLES				
SIGNALLING CABLES (SEE NOTE)				
CATENARY MESSENGER & CONTACT WIRES (SEE NOTE)				

ASSUMPTIONS :-

1. FE EQUIPMENT AND MATERIALS DELIVERED 1 MONTH BEFORE INSTALLATION

2. COOLING PIPE WILL BE FABRICATED OUTSIDE THE SOUTH CHERITON PORTAL AND STORED IN 3 LOCATIONS IN THE RUNNING TUNNELS

3. TEMPORARY CROSSOVERS WILL BE CONSTRUCTED AS FOLLOWS :-
    - ST-RTN - CH.36.6
    - ST-RTS - CH.37.1

4. INTERMITTENT TRAIN TESTS IN LRTS MAY/JUNE 1992

5. HIGH SPEED TRAIN TEST IN RTN TO XOVER 31-OCT/13-NOV-92 WITH ALL TUNNELS EVACUATED

NOTE :-

1. REGISTRATION AND FINAL EARTHING & BONDING OF CATENARY MESSENGER AND CONTACT WIRES IS NOT INCLUDED

2. SIGNALLING BOXES ON TRACK NOT INCLUDED

3. HANDRAILING NOT INCLUDED

DRAWING No : B10/100/8/IPO/207/PG/VLO
DATA DATE : 31-MAR-91
ISSUE DATE : 04-AUG-91
REVISION No: C

SHEET 1 OF 6

LAYERS 1,18,23,26 & 27

DRAFIX FILE : TUNFEMA1.DWG

1991

MAY	JUN	JUL	AUG	SEP	OCT	NOV	DEC
12-May	09-Jun	07-Jul	04-Aug	01-Sep	29-Sep	27-Oct	24-Nov	22-Dec

ch 10.000
ch 15.000
ch 20.000
ch 25.000
ch 30.000
ch 35.000
ch 40.000
ch 45.000

CASTLE HILL/HOLYWELL

EARTHING BONDING & INSTRUMENTATION

HYDROTEST/EPOXY COATING

PIPE STORAGE

MARSH AREA

EARTHING BONDING & INSTRUMENTATION

HYDROTEST/EPOXY COATING

CROSSOVER

PIPE STORAGE

XOVER

SANTRES

TBM BURIAL

EXTRACT FROM
28 MONTH
PROGRAMME

12-May	09-Jun	07-Jul	04-Aug	01-Sep	29-Sep	27-Oct	24-Nov	22-Dec
MAY	JUN	JUL	AUG	SEP	OCT	NOV	DEC	

1991

Fig. 30. Extract from level 2 time chainage programme: running tunnel north linear

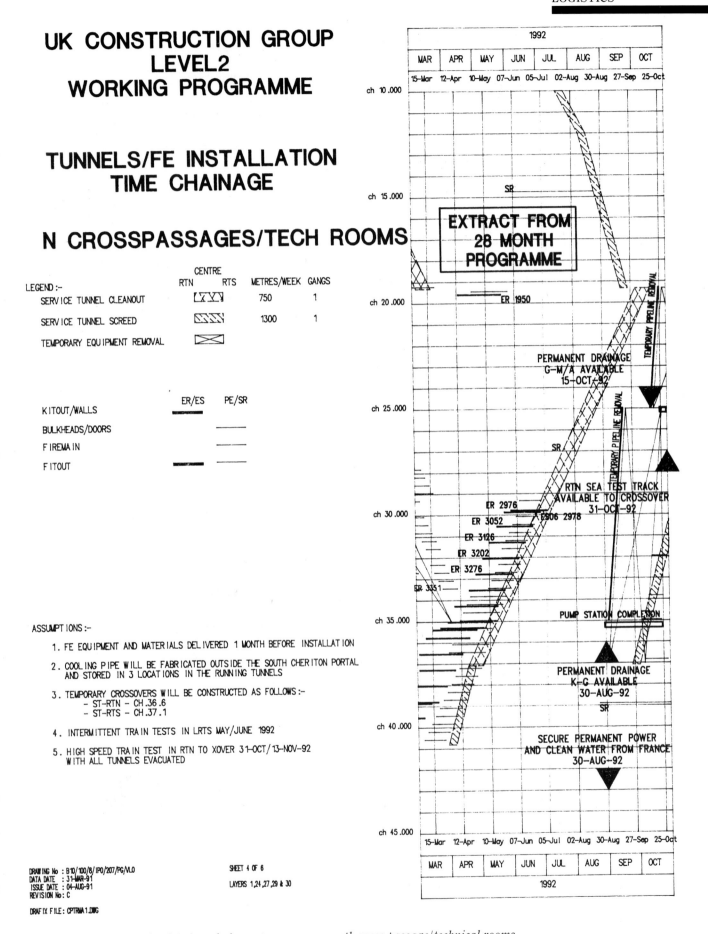

UK CONSTRUCTION GROUP
LEVEL 2
WORKING PROGRAMME

TUNNELS/FE INSTALLATION
TIME CHAINAGE

N CROSSPASSAGES/TECH ROOMS

LEGEND :-

	CENTRE		METRES/WEEK	GANGS
	RTN	RTS		
SERVICE TUNNEL CLEANOUT			750	1
SERVICE TUNNEL SCREED			1300	1
TEMPORARY EQUIPMENT REMOVAL				

	ER/ES	PE/SR
KITOUT/WALLS		
BULKHEADS/DOORS		
FIREMAIN		
FITOUT		

ASSUMPTIONS :-

1. FE EQUIPMENT AND MATERIALS DELIVERED 1 MONTH BEFORE INSTALLATION

2. COOLING PIPE WILL BE FABRICATED OUTSIDE THE SOUTH CHERITON PORTAL AND STORED IN 3 LOCATIONS IN THE RUNNING TUNNELS

3. TEMPORARY CROSSOVERS WILL BE CONSTRUCTED AS FOLLOWS :-
   - ST-RTN - CH.36.6
   - ST-RTS - CH.37.1

4. INTERMITTENT TRAIN TESTS IN LRTS MAY/JUNE 1992

5. HIGH SPEED TRAIN TEST IN RTN TO XOVER 31-OCT/13-NOV-92 WITH ALL TUNNELS EVACUATED

DRAWING No : B10/100/8/IPO/207/PG/VLO
DATA DATE : 31-MAR-91
ISSUE DATE : 04-AUG-91
REVISION No : C

DRAFIX FILE: OPTRWA1.DWG

SHEET 4 OF 6

LAYERS 1,24,27,29 & 30

Fig. 31. Extract from level 2 time chainage programme: north cross-passage/technical rooms

aged the auditee to volunteer how he might work more safely and productively; these reinforced the objective of running a safe, orderly and productive railway, to which managers, supervisors and operatives all contributed.

Planning following completion of tunnel drives

84. During the tunnel driving the overall strategic priorities were quite clear and the interfaces with our French colleagues obvious. Once tunnel driving had been completed the picture was more complex. However, the planning methods used for tunnel construction proved capable of further development.

85. Strategic planning at level 2 was carried out by means of time chainage programmes for individual tunnels, and these were linked to similar programmes in France. A suite of six time chainage programmes was produced which showed all fixed equipment installation activities and the temporary services strip-out (see Figs 29–31). This suite of programmes contained the following.

(a) Cleanout and trackwork: combined RTN and RTS to show the relationship of these capital plant intensive activities and to form the basis of the other more flexible operations.
(b) RTN—linear activities.
(c) RTS—linear activities: showing running tunnel cleanout and trackwork plus pre-track and post-track cabling and mechanical installations.
(d) ST linear activities: showing cleanout, linear cabling and mechanical activities and indicating major structures.
(e) South cross-passages and technical rooms.
(f) North cross-passages and technical rooms: showing all phases of technical room installation, temporary services removal, permanent services availability and service tunnel cleanout.

86. The programmes were accompanied by schematics clarifying the activity descriptions, the UK/France interface and the changing phases of tunnel access. Detailed planning was then carried out on the level 3 time chainage format developed during tunnelling (see Fig. 32). These included all sub-contractor activities and extended for the whole tunnel with a 6 month time span. The weekly four-week rolling programmes were also continued, with sub-contractors now included in the co-ordination process.

Planning systems

87. The project-wide level 2 networks and the transportation system engineering level 3 design and procurement networks were based on Artemis 9000 software, and were processed on TMLs networked IBM ES 9000 mainframe computer systems. The sub-project level 3 and 2 networks utilized Artemis 7000 mounted on IBM PS2 series 70 (386) PCs. The level 2 time chainages were drawn using Drafix mounted on IBM Series 50 PCs.

Project management control and progress monitoring

88. Having developed sophisticated suites of programmes, it was necessary to manage the work in accordance with them. During tunnelling, progress was easily monitored; in the post-tunnelling phase the numerous activities were spread over nearly 84 km of tunnels, and hence progress was not so easily measured. From the time chainages a comprehensive level 2 network was created. Beneath this 37 level 3 networked bar charts were produced, covering each major structure and system, and each rolling up to the relevant level 2 activities. A series of weekly reports were produced to show planned and actual progress, in the form of percentages, bar charts and S-curves, together with the progress trends (see Fig. 33).

89. Reports were produced for overall progress, progress by sub-contractor and progress by system. From these the sub-project management knew where to concentrate their efforts to complete the work in accordance with the overall strategic plan.

90. The level 2 sub-project network was integrated into a project-wide network and progress reports produced in this wider context.

Conclusion

91. When system commissioning commenced in early 1992, the tunnel was divided into 14 commissioning zones. Power system commissioning commenced at the portals working towards mid-Channel. The planning systems successfully developed during tunnelling were continued into this phase of work. The operation of standard gauge trains was controlled in a similar manner to that successfully used for the narrow gauge railway system, as were the rubber-tyred vehicles for temporary use, which eventually supplanted the narrow gauge railway in the service tunnel.

92. The hierarchical planning arrangement is not uncommon, and its use proved eminently suitable for the Channel Tunnel project. However, on a fast-track project it must be accepted that level 4 will hardly ever roll up to level 3, and that level 3 will not always roll up to level 2. A pragmatic approach to interfacing must be adopted.

93. The level 4 four-week look-ahead programme was an admirable demonstration that the key to successful co-ordination is not the piece of paper which is produced but the discussions and decisions leading to it. This should be true for all planning processes.

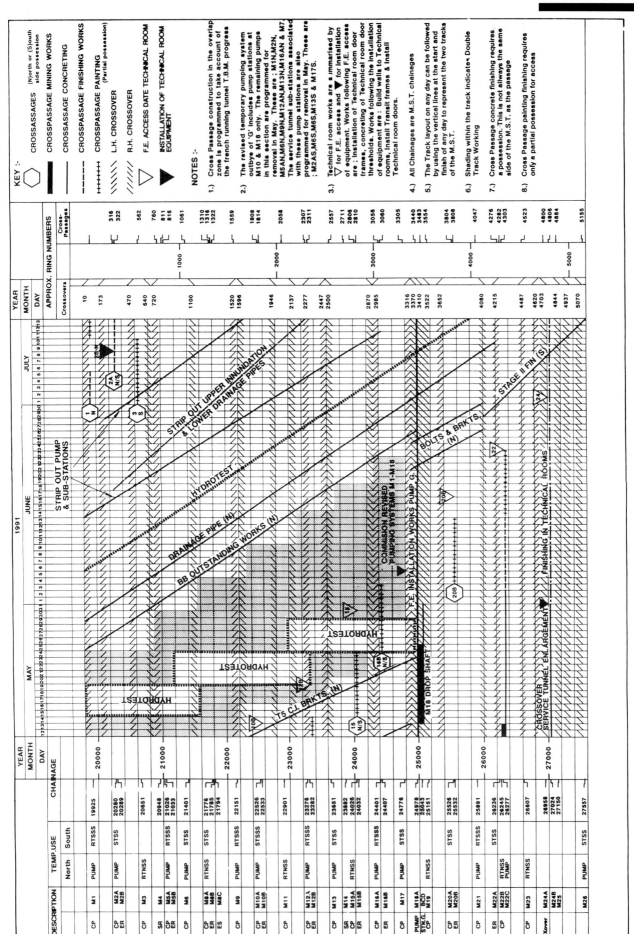

Fig. 32. Extract from 6 month level 3 time chainage programme (installation phase)

125

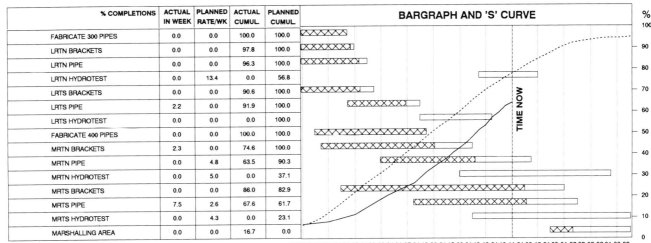

TUNNEL CONSTRUCTION - WEEKLY SUMMARY REPORT **WEEK ENDING 10-11-91**

COOLING SYSTEM

% COMPLETIONS	ACTUAL IN WEEK	PLANNED RATE/WK	ACTUAL CUMUL.	PLANNED CUMUL.
FABRICATE 300 PIPES	0.0	0.0	100.0	100.0
LRTN BRACKETS	0.0	0.0	97.8	100.0
LRTN PIPE	0.0	0.0	96.3	100.0
LRTN HYDROTEST	0.0	13.4	0.0	56.8
LRTS BRACKETS	0.0	0.0	90.6	100.0
LRTS PIPE	2.2	0.0	91.9	100.0
LRTS HYDROTEST	0.0	0.0	0.0	100.0
FABRICATE 400 PIPES	0.0	0.0	100.0	100.0
MRTN BRACKETS	2.3	0.0	74.6	100.0
MRTN PIPE	0.0	4.8	63.5	90.3
MRTN HYDROTEST	0.0	5.0	0.0	37.1
MRTS BRACKETS	0.0	0.0	86.0	82.9
MRTS PIPE	7.5	2.6	67.6	61.7
MRTS HYDROTEST	0.0	4.3	0.0	23.1
MARSHALLING AREA	0.0	0.0	16.7	0.0

BARGRAPH AND 'S' CURVE

31-03-91 28-04-91 26-05-91 23-06-91 21-07-91 18-08-91 15-09-91 13-10-91 10-11-91 08-12-91 05-01-92 02-02-92 01-03-92

OVERALL SUMMARY

STATUS -6.5 WEEKS

	ACTUAL %	PLANNED %
IN WEEK	1.67	1.75
CUMUL.	64.12	77.83

NOTES

1. BASED ON LEVEL 2 PROGRAMME REV 6A

PRODUCED BY UK TUNNELS PLANNING DEPARTMENT

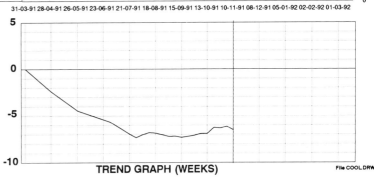

TREND GRAPH (WEEKS) File COOL.DRW

Fig. 33. Typical weekly summary sheet

Acknowledgements

94.   The development of the Lower Site, the operation of the pithead and the operation of the construction railway were all vital to the success of the tunnel drives. The people involved ranged from divers, drivers, planners, computer programmers, construction managers, engineers of all disciplines, canteen staff and tunnel 'hygiene' crews. Without the dedication of all concerned, the great success of this work would not have been possible. The Authors are grateful to Eurotunnel and TML for permission to publish this Paper.

Reference

1.   POLLARD C. and KERSHAW K. UK Channel Tunnel drives construction and environmental aspects of the disposal of tunnel spoil. *Proc. conf. recycling of construction and demolition waste.* ENTSORGA '91, Essen.

Tunnel lining design and procurement

R. C. W. Eves, BSc(Eng), MICE and D. J. Curtis, BSc(Eng), MSc(Eng), DIC, FICE

Proc. Instn Civ. Engrs, Civ. Engng, Channel Tunnel Part 1: Tunnels 1992, 127–143

Paper 10045

■ In this Paper the design studies and methods are described for both precast concrete and spheroidal graphite cast iron (SGI) linings, together with an outline of the design requirements of the construction contract. The Paper also gives a brief overview of the UK precast concrete manufacturing facility, where 442 755 reinforced concrete tunnel lining segments were produced. It describes the concrete mix development, the manufacturing processes and the systems by which the quality of lining was assured. A brief overview is given of the SGI lining requirements and the procurement of 38 000 tonnes of linings, which required a worldwide search for the manufacturing capacity.

Introduction

The preformed tunnel linings described in this Paper provide support to the ground in the excavations carried out from the UK in the full-face tunnel boring machine (TBM) drives, and in the cross-passages and adits themselves.

2. The six UK tunnel boring machine (TBM) drives, for the two running tunnels of 7·6 m internal diameter and the 4·8 m internal diameter service tunnel (three underland towards the UK terminal from Shakespeare Cliff and three from the Cliff towards France) required 81·9 km of preformed segmental tunnel linings. There are 306 cross-passages, interconnecting the service and running tunnels for passenger evacuation and the accommodation of mechanical and electrical equipment, including pumping stations. These mostly have 3·3 m internal diameters, but a proportion are 4·0 m and 4·8 m. In addition, over one-hundred 2·0 m internal diameter pressure relief ducts connect the running tunnels and pass over the service tunnel at 250 m centres. In all some 800 junctions between tunnels have been constructed, and special preformed linings have been designed and erected in the main tunnels at these locations.

3. The preformed linings have been fabricated in reinforced concrete or spheroidal graphite cast iron (SGI) segments, although a limited number of the openings in the main tunnels have incorporated structural steelwork. The first part of this Paper describes the design of the linings. The two subsequent parts describe the manufacture of the precast concrete linings, which was undertaken by Transmanche Link (TML), and the procurement of the cast iron linings.

4. The final part of the Paper draws some conclusions from the experience gained from the implementation of the project. At the time of writing the tunnelling works have been successfully completed within the programme envisaged when TML were awarded the contract in May 1986.

Design

5. When the concession for the Channel Tunnel project was awarded in 1986 the construction programme called for very early definition of the requirements for the tunnel boring machines and for the procurement of the prefabricated segmental tunnel linings in both reinforced concrete and spheroidal graphite cast iron. The experience gained in the aborted project of the 1970s[1,2] led to the conclusion that the excavation of the main tunnel drives should, for economy and speed, be carried out by fully automated machines, and that the tunnels should be lined with preformed structural elements erected immediately behind the machines. References 3–6 contain a considerable amount of information upon the development of the lining designs. It is not intended to restate all that has already been published, but to place in fuller context the constraints and possibilities for design development that existed in the present 'fast-track' design and construct project.

6. Under the contract the design was required to satisfy the following provisions.

7. 'All of the permanent works are to be designed for a life of 120 years. Such items as fixings, and internal caulking shall have a design life of 25 years and shall be accessible.'

8. 'Loadings—design model. The calculations shall conform to two separate limit states. The first being the standard design cases taking account of the interaction between the lining and the ground, the in situ state of stress in the ground, the deflection of the linings and the redistribution of the loading dependent upon the relative flexibility and compressibility of the lining. The second being to ensure that the ultimate limit state of the lining is checked against the unfactored full ground and water overburden.'

9. 'An appropriate allowance for seismic loading shall be required.'

10. 'Watertightness. There shall be no dripper in the upper half of the tunnel onto sensitive equipment (not more than one drip per minute). There shall be no continuous leaks (more than 4 litres/hour) other than those being diverted directly into the drainage system.'

R. C. W. Eves, Construction Manager, Isle of Grain Precast Site, Transmanche Link

D. J. Curtis, Director, Tunnel Division, Mott MacDonald

Fig. 1. General view of running tunnel showing precast concrete and hybrid concrete cast iron linings

Fig. 2. Piston relief duct SGI lining with erection formers in place

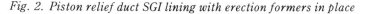

11. 'The maximum value of lack of circularity, due to both movement underground as a result of loads and building tolerance, shall not exceed 1% of the radius.'

12. The contract also contained a list of mainly French and British design codes and standards. None of these documents was drawn up while taking into account the special circumstances which might be found in their application to tunnels, and they included a mixture of 'limit state' and 'working stress' codes. The French and British designers were, however, in agreement that the inclusion of the list imposed upon them the need to justify their designs according to the best standards expected of them. Limit state methods were usually adopted.

13. It is interesting to reflect upon the five design provisions contained in the contract and summarized above. The first contains two design lives, differentiating between 'permanent works' and 'fixings and internal caulking which shall be accessible'. A design life can be defined in terms of fatigue, durability by means of deterioration of material properties by chemical changes, and the effects of extraordinary physical events of which only accidents and earthquakes are relevant to tunnels. Fatigue is not a significant feature for a tunnel lining which is designed to work in a mainly compressive permanent load regime. However, it has to be considered carefully in terms of the trackbed structure supporting the rails, which could affect the lining itself if the ground surrounding the tunnel were to be sufficiently soft to be susceptible to liquefaction under repeated low-stress fluctuations. The latter consideration does not apply in this project.

14. The second provision clearly demands that a 'limit state' design philosophy be adopted. Usually the two separate limit states are 'serviceability' and 'ultimate'. In the contract the first limit state can be taken to involve both of these, but the second is an alternative ultimate limit state, which at first glance would always govern the design since the 'interaction' analysis leads to ground loads of only about one-half of the overburden. It was, however, established that in the second case it would be assumed that the linings had been perfectly constructed, and for the undersea tunnels this case rarely governed. The application of the second case to the UK underland tunnels can be held to have resulted in unnecessarily thick and heavy segmental linings. The application of this rule to non-circular tunnels took longer to establish.

15. Some time elapsed before consensus upon seismic events could be reached. In the meantime it was established that underground structures could readily withstand ground accelerations of 0·05 g (serviceability) and 0·2 g

(ultimate), which were recognized as being reasonable seismic loading estimates. Only certain structures built upon shafts required more rigorous investigation. Within the underground structures care was taken to provide for movements at major changes of dimension, such as at the entry of the running tunnels into the crossover caverns.

16. There is some difficulty in reconciling the fourth and the first provisions as far as durability is concerned in the undersea tunnels, since the contract did not define the acceptable frequency of occurrence of drippers and continuous leaks. This is especially significant in view of the potential for spreading harmful chloride-rich inflow waters along the tunnel by the operation of the trains in the railway system. The specification subsequently developed for the tunnels called for a higher standard of inflow water control than could have been interpreted from the fourth provision.

17. The fifth provision was considered reasonable from an examination of records from soft ground tunnels previously constructed.

18. Under the terms of the contract, TML were required to produce a design manual for the detailed design stage of the project. Development studies were called for in advance of the manual in order to resolve uncertainties in the design methods for tunnel linings and structures. The tunnel lining development study reviewed loadings upon tunnel linings in view of theoretical advances as well as evidence obtained from earlier projects, especially that of the project abandoned in 1974.[2] Tests and investigations were also started into load transfer across radial joints in rings formed from segments in precast concrete and cast iron, and into waterproofing and seepage water inflow control measures.

19. Load cases and imposed deformations are listed in Reference 3; this also lists the partial safety factors adopted in the limit state calculations implied in the construction contract. These recognize the predominance of predeterminable ground and water loadings upon the tunnel linings, and that circular tunnels are more stable than most structures, since they act mainly in compression and are far less affected by bending moments than are beams and columns in other more common structures. This is supported by Reference 7.

20. Surveys of the ground through which the tunnel might be driven have been carried out since 1832 and culminated in the supplementary geotechnical and geophysical investigations for the present project[8] in 1986–87. The tunnels are mainly located in the favourable clayey but cemented Lower Chalk Marl, but must sometimes pass through or close to adjacent strata (Fig. 3).

21. The engineering properties of the Lower Chalk Marl vary considerably over short depths

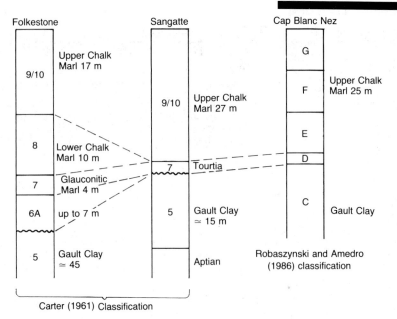

Fig. 3. Simplified cross-channel stratigraphical correlation

Fig. 4. Variation in engineering properties of Chalk Marl

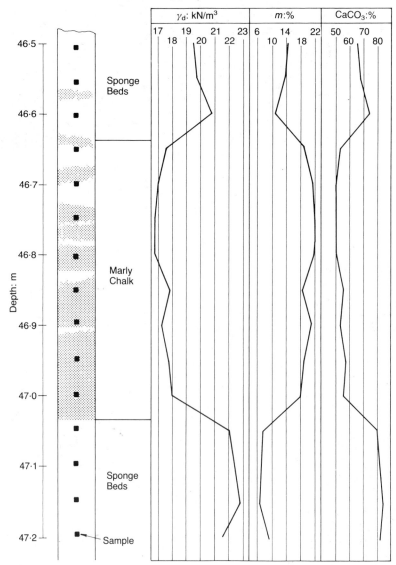

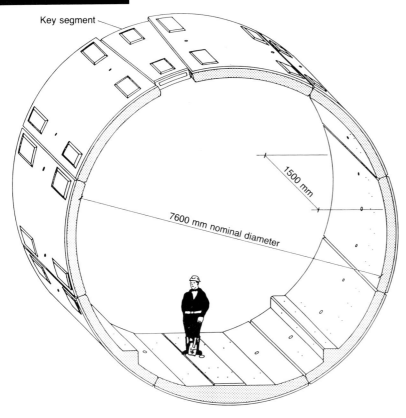

Key segment

1500 mm

7600 mm nominal diameter

Fig. 5. Typical precast concrete running tunnel lining

(*b*) the creep of the ground, including all time-dependent effects;

(*c*) horizontal ground pressures as a proportion of overburden pressures.

Back analysis of the measurements reported in Reference 2 provides a check on the values of the parameters: Young's modulus = 760–1400 MPa, creep ratio = 1·0–1·6, and effective ratio of pre-existing ground stresses = 0·7.

22.   A list of values of parameters adopted in the design of the tunnel linings in the first part of the UK undersea drives is given in Table 1. It must be remembered that the values assumed are also dependent on the method and form of construction. For example, the 'effective' ratio of pre-existing ground stresses is not the value measured during in situ tests, which would indicate values often in excess of unity. Values closer to those measured were employed in calculations for the shotcrete linings in those parts of the projects constructed by the New Austrian Tunnelling Method.

23.   The design method for the linings described in this Paper has as its simplest expression the Kelvin visco-elastic model. For uniform radial loads this can be expressed as

$$N = [P_0]\left[(1 - \lambda) + \lambda\left(e^{-\gamma T_0}\frac{\Phi}{1 + \Phi}\right)\right]\left[\frac{1}{1 + Q}\right][R] \quad (1)$$

where N is the hoop load acting in the lining, λ is the deconfinement ratio, γ represents a rate of creep of the ground, Φ is the ratio of creep to immediate (elastic) strains in the ground, T_0 is

(Fig. 4) and fissures are to be expected. The values for geotechnical parameters assumed from small-scale tests are modified to represent more accurately the actual rock mass behaviour, using information gathered from earlier projects. The predominant parameters are

(*a*)   rock mass values of Young's modulus;

Table 1. *Parameters adopted in design of first part of UK undersea drive*

Parameter	White chalk	Grey chalk	Upper chalk marl	Stratum lower chalk marl	Glauconitic marl	Tourtia	Gault clay
γ_{sat}: kN/m³	21·5	22·5	22·5	23·0	23·5	22·0	21·5
UCS: MPa		9·0	9·0	6·0	4·7	4·0	3·0
E		1300	1300	900	900	500	250
E_{50}		975	975	675	600	375	175
Φ (creep)		1·5	1·5	1·5	1·5	2·1	2·1
v (drained)		0·3	0·3	0·3	0·3	0·3	0·3
v (undrained)		0·5	0·5	0·5	0·5	0·5	0·5
Pre-peak							
c': MPa		1·2	1·2	0·9	0·5	0·5	0·25
ϕ': deg		35	35	35	30	30	30
Post-peak							
c': MPa		0	0	0	0	0	0
ϕ': deg		35	35	35	30	30	20
K_0 (longitudinal)		1·5	1·5	1·5	1·5	1·3	1·3
K_0 (transverse)		0·5	0·5	0·5	0·5	0·7	0·7
Mass permeability: m/s		1×10^{-9}	2×10^{-7}	7×10^{-8}	7×10^{-8}	1×10^{-9}	

the time delay between excavation and lining, P_0 is the pre-existing radial stress in the ground, and Q is the ratio between the elastic constants of the ground and lining (E, v), modified for creep effects, multiplied by the radius R divided by the lining thickness.

24. This model can be extended to anisotropic stress fields and to plastic ground behaviour;[9,10] this has been done in the detailed design stage. In the above formula, values for λ were taken as 1·0 for the UK TBM drives and 0·6 for cross-passages. The value of the term $e^{-\gamma T_0}$ was usually assumed to be 1·0.

25. For watertight linings the calculation of ground load is made where effective stresses and water loading are computed separately from the relationship

$$N_w = [P_w]\left[\frac{1}{1+Q}\right][R] \qquad (2)$$

The precast linings are not watertight and water pressures are dissipated and transferred to the surrounding ground. As a first approximation, the ground and water loads can be combined and their effects calculated using equation (1).

26. Finite-element analyses were carried out in parametric studies to establish the importance of various parameters upon the designs. In general two-dimensional analyses incorporated visco-elastic and/or visco-plastic models, whereas the three-dimensional analyses required for the design of openings and junctions usually used a linear elastic model. Studies were carried out to determine *inter alia* the effects of

(a) ground strata interfaces;
(b) subsequent tunnels upon the first;
(c) a running tunnel on a previously constructed cross-passage, etc.;
(d) junctions upon the main tunnel linings.

It was possible in most situations to draw up simple rules with which to design the linings at any location, for example: hoop loads are those due to the weaker of those present at an interface which touches or intersects the excavations; the running tunnels will increase loads in the pre-existing service tunnel by 20% under-sea and 25% underland. The hoop loads can then be determined from the deterministic formulae contained in Reference 4 using the values of ground parameters adopted for this method.

27. For economy, precast concrete (PCC) segmental linings were the preferred choice for the majority of the main tunnel drives. In the UK drives, articulated and expanded linings were used (Fig. 5). In the French drives, articulated and bolted linings were adopted to suit the predominant wet and fissured ground conditions. The logic behind these choices has already been discussed[4] and has been justified in the event.[6]

28. In developing the design of the PCC linings, the scale of the project, the need for rapid progress both in establishing the supply of segments and during construction in the tunnel, and an obligation to use the best current (1986) practice in workmanship and materials were all taken into consideration. The outcome, shown typically in Figs 5 and 6, incorporated large segments 1·5 m long, over 3 m in arc length, and varying in thickness from 270 to

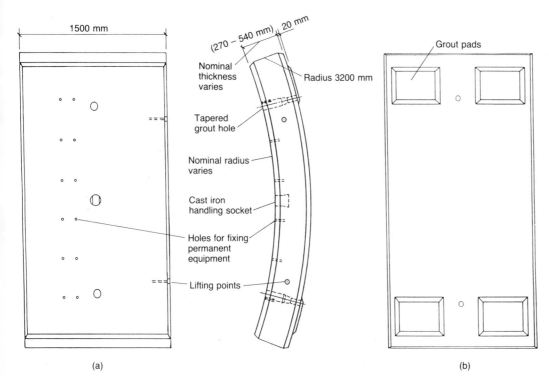

(a)

(b)

Fig. 6. Typical arrangement of UK precast concrete tunnel lining segment: (a) front elevation; (b) rear elevation

540 mm. The details shown in Fig. 6 indicate some of the ways in which the precasting facility was used to provide features to simplify handling and to speed construction. For example, the integral grout pads to aid ring building, and to ensure a continuous path for cavity grouting, posed a technical challenge for the supplier. The holes for fixing permanent equipment are another example of the intentions to pursue the one-pass lining approach as far as possible. Other details not shown in the figure includes rebates at the ends of mating circumferential and radial joint surfaces to minimize the risk of damage during construction in the tunnel.

29. It was necessary to provide steel reinforcement in such large segments primarily for safety during handling and erection, and to resist tensile bursting stresses near the articulated radial joints which cause concentrations of the compressive hoop load in the rings. This in turn raised the problem of durability due to corrosion of the reinforcement, which was a serious concern in view of the saline nature of the ground water expected in the undersea tunnels. In order to control the corrosion problem the following measures were taken.

Fig. 7. Comparison of concrete mixes showing the Channel Tunnel concrete to have exceptionally low permeability and diffusibility

(a) A concrete mix of exceptionally low permeability and diffusibility was provided, achieved as illustrated in Fig. 7.

(b) Steel reinforcement was covered to 35 mm, a compromise between the conflicting demands of durability and strength of the segments, which is compromised if the cover/thickness ratio is too large.

(c) The cavity grouting of the annulus between the lining and the excavated ground, together with proof-grouting as necesary to minimize water inflows.

(d) Closed drainage paths were constructed for such water that may penetrate the joints between segments.

(e) Reinforcing cages were fabricated by welding, and were provided with bonding terminals for possible future localized cathodic protection schemes.

30. Permeability and diffusivity tests were carried out during the mix development, and during production. In view of the quality of the concrete, the tests took a long time to carry out, and the advancing chloride front was detected by drilling holes and testing the powder produced. A computer program was developed[11] and employed to estimate the time at which critical concentrations of chloride would reach the reinforcing bars.

31. The design of the curved radial joints has been discussed in Reference 4. Reinforcement is provided to suit the design load as derived from small- and large-scale laboratory tests, employing the design chart shown in Fig. 8. Also shown are the comparable design rules included in British Standard 8110.

32. The load-carrying capacity of precast concrete linings was adjusted by the amount of steel reinforcement included. At junctions with cross-passages and other areas of locally enhanced loads, ductile SGI was employed in

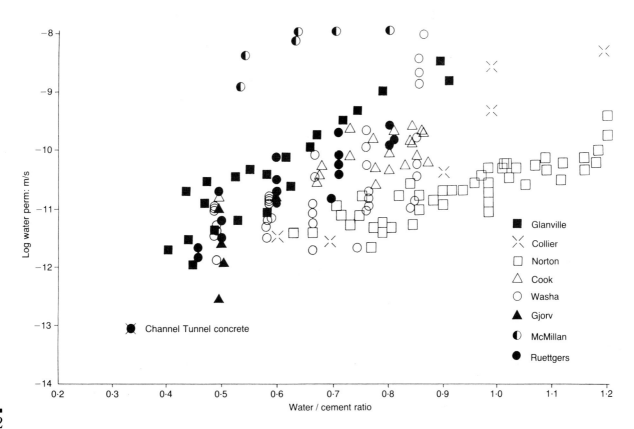

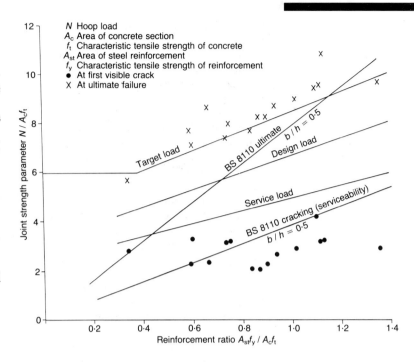

Fig. 8. *Design chart compiled from joint test results*

order to avoid needing to increase the tunnel size everywhere to accommodate such loads. In the UK drives it was also necessary to provide alternative bolted SGI linings for ground conditions in which an articulated expanded concrete lining could not be built to satisfactory standards. Although there have been a few previous applications in tunnels of SGI those have mainly been based upon the substitution of SGI for the brittle Grey Iron used over the previous hundred years. The tensile strength and ductility advantages of SGI had rarely been used. A research project undertaken in the mid-1980s[12] showed the potential for cost savings, and further tests commissioned by TML served to establish sensible design parameters. These were adoped in the project, but, except for hand-mined tunnels such as cross-passages and pressure relief ducts between the main TBM drives, economics dictated that the use of SGI should be minimized in the main drives. It was also found that the casting process did not allow the full material economy of the design calculation to be achieved when casting tolerances were very large in terms of design thicknesses. Comparisons between SGI and high tensile steel are shown in Fig. 10.

33. The design rules adopted are described in Reference 4. In order to guard against corrosion a 1 mm allowance was added to the thicknesses of the skin and peripheral circular flanges of cast iron segments. This does not necessarily provide for 120 years of deterioration, but it gives time to discover deterioration and to take measures to counteract it. All surfaces were coated with a bituminous paint.

34. The appearance of the bolted cast iron linings is shown in Fig. 19. The rings designed for the main tunnels were built using segments which could be man-handled if necessary. For mechanical erection, panels of four segments were bolted together on the surface and then transported and erected as one piece. All flanges were machined and each incorporated a groove to accommodate a rubber gasket (EPDM) if required. Linings designed for use in hand-mined tunnels were generally provided with a smaller groove into which a hydrophilic gasket could be inserted.

35. Initially it was anticipated that at junctions between the main tunnels and cross-passages and ducts, several rings of SGI linings would be placed in the main tunnels. The original concept was that rings forming openings in the main tunnels should be built one-by-one without interruption (Fig. 11). TML in the UK developed this concept further, whereby the construction and erection of all segments should not differ, whether or not their principal material was concrete or cast iron. From this the 'hybrid ring' was developed (Fig. 12). In the hybrid openings the use of SGI segments was

Fig. 9. *Reinforcement cage for 540 mm thick concrete segments*

Fig. 10. *Stress versus strain for grade 600/3 SGI and grade 50 structural steel*

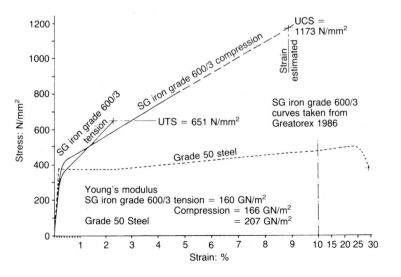

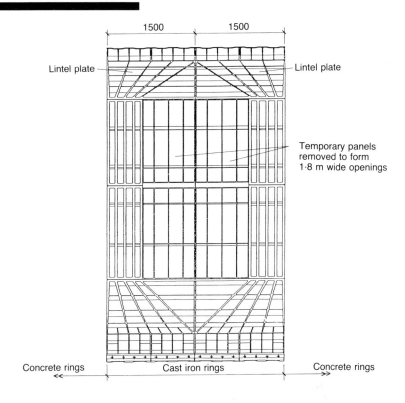

Fig. 11. Cross-passage opening set in service tunnel

Fig. 12. Elevation of running tunnel hybrid opening set

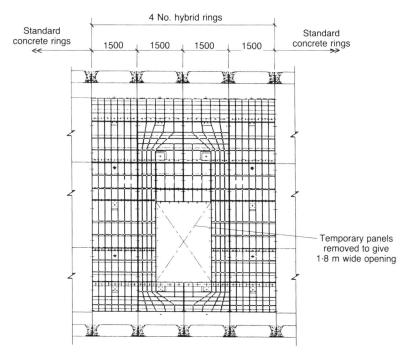

limited to segments immediately adjacent to the openings.

36. Pressure relief ducts between the main train running tunnels have to pass over the intermediate service tunnel at regular (250 m) intervals. For aerodynamic reasons these tunnels need a smooth internal bore, and from construction considerations a lightweight SGI lining is desirable. TML proposed a lining in which the circumferential skin was entirely in the intrados and required an elaborate system of temporary connections (Fig. 13). Despite the fact they were built in difficult circumstances, construction of the ducts has proceeded with great success.

37. In all some 15 different rings of concrete lining were designed, 16 rings of cast iron and 12 designs of opening sets.

Precast concrete lining manufacture

Overview of the Isle of Grain precast site

38. For the UK tunnels 442 755 reinforced precast concrete lining segments were required to be manufactured within a $3\frac{1}{2}$ year period. A suitable site was identified at the Isle of Grain in the north of Kent. The location was chosen because it provided an area of 29 ha, which was environmentally acceptable, and had existing deep-water dock facilities with road and rail access.

39. Work to transform the barren marsh land close to the Thames Estuary to establish the largest precast facility in Europe, at that time, commenced in November 1986 (Figs 14 and 15).

40. The specification required that the best manufacturing techniques and materials were used for the linings. The production lines and moulds were provided by Sacma Spa (Milan) and the 1 000 000 tonnes of fine and coarse granite aggregate were shipped from Foster Yeoman's quarry in Scotland.

41. The 200 000 tonnes of Blue Circle cement were transported by rail from their works at Northfleet, and the 90 000 tonnes of pulverized fuel ash and 45 000 tonnes of small diameter reinforcement were delivered by road transport from Ash Resources works at St Neots and ROM Ltd's reinforcement works at Lichfield, respectively.

Concrete mix development

42. Investigations were carried out on various aggregates and combinations of Portland cement, cement replacement materials (GGBFS and PFA) and admixtures. The chosen materials were selected for operational/technical performance, consistency of supply and commercial considerations. From extensive laboratory trials the following initial mix was developed, and it was determined that a curing regime of 1 h at 20°C followed by 5 h at 50°C satisfied both the specification and the proposed automated production method.

OPC	310 kg/m³
PFA	130 kg/m³
20 mm granite	818 kg/m³
10 mm granite	358 kg/m³
Processed granite fines	638 kg/m³
w/c ratio	0·35
Superplasticizer (Sikament FF)	5·0 l/m³

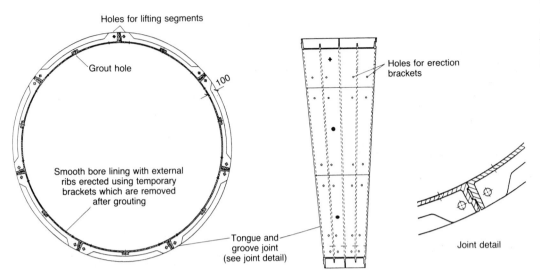

Fig. 13. 2 m 1D pressure relief duct linings ring length 0·75 m

Fig. 14. Aerial view of Isle of Grain precast site

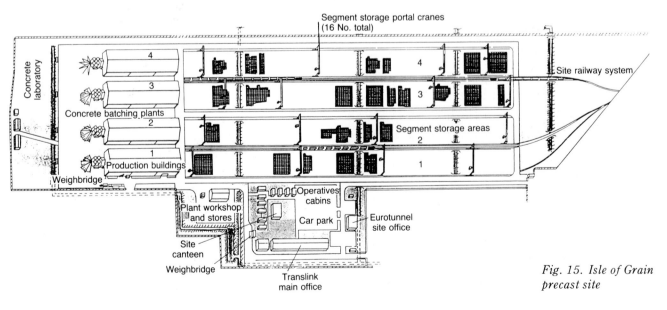

Fig. 15. Isle of Grain precast site

135

43. The first production line was successfully commissioned in late September 1987 and production commenced on 12 October 1987. During the early production phase two particular problems were encountered with the concrete mix. These related to voids occurring below the reinforcement, and the production of a satisfactory surface finish to the moulded faces for the durability requirements. The problem of voids under the reinforcement was overcome by modifying the aggregate proportions by increasing the small grit and reducing the coarse zone, thereby increasing the mechanical interlock and the total aggregate surface in relation to the cement paste, together with an examination of the method of vibration/compaction.

44. The problem of surface blemishes was resolved after considerable research by British Petroleum, who developed a specific solvent-refined paraffinic mould oil that satisfied both health and technical requirements.

45. The final mix that was used throughout production is shown below. It was compacted using a 65 mm diameter Whacker electric motor in the head poker vibrators, which were electronically synchronized with the rate of concrete placing/withdrawal.

OPC	310 kg/m³
PFA	130 kg/m³
20 mm granite	580 kg/m³
10 mm granite	454 kg/m³
Processed granite fines	780 kg/m³
w/c ratio	0·35
Superplasticizer (Sikament FF)	5·8 l/m³

The specification required a minimum concrete strength of 60 N/mm², and the typical strength achieved for the production concrete was 70–75 N/mm² at 28 days, with a further gain to 90–95 N/mm² at 90 days. In order to control fully the production process, a method of assessing the effect of the following variables on the concrete needed to be established

(a) initial mix temperature;
(b) thickness of segment;
(c) curing tunnel temperature;
(d) process (cycle) time;
(e) temperature differentials throughout the segment.

46. After much investigation the Danish Concrete and Structural Research Institute's Computer Integrated Test and Monitoring System (CIMS) was found to be the most accurate method of control. The CIMS concept was developed in Denmark in the 1960s by Professor Freiesleben Hansen. At the Betonog Konstruktion Instituttet (BKI), it was further developed into a computerized planning and control system that could be sold commercially.

47. The system was programmed with the determined adiabatic calorimetry and strength development data obtained from the production concrete mix. Using this information as a base, by inputting or changing the variable parameters, the strength, maturity and temperature differential at any stage could be predicted. Verification and correlation of the system was carried out for each type of segment by the use of thermocouples connected to a comprehensive data logger. The simulation system compared very favourably with the actual monitoring, and thus greatly reduced the 'reaction time' required while monitoring day-to-day production when one or more of the variables changed.

48. The number of variable combinations are too numerous to list; however, the main effects were due to seasonal temperature changes. This resulted in cycle times varying between 7½ min for 270 mm thick segments when the initial mix temperature could be maintained at 25°C, to 15 min for 540 mm thick segments with an initial mix temperature of 15°C.

Fabrication of reinforcement cages

49. The supply and manufacture contract for the reinforcement cages was awarded to ROM Ltd, who manufactured the piece parts at their main factory at Lichfield and fabricated completed cages at their Isle of Grain factory. A typical cage is shown in Figs 16 and 9. At ROM's Lichfield factory CARES-approved steel was used to produce the cage piece parts. The ladder mats were manufactured from 12 mm hard drawn wires which were connected by resistance welding. A minimum shear strength

Fig. 16. Typical tunnel lining segment reinforcement: (a) section C-C chair; (b) plan; (c) section D-D ladder; (d) elevation

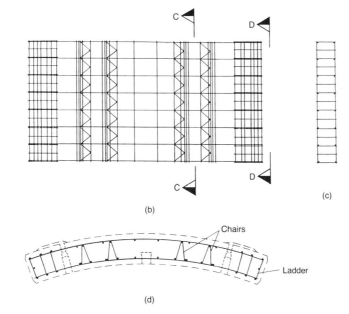

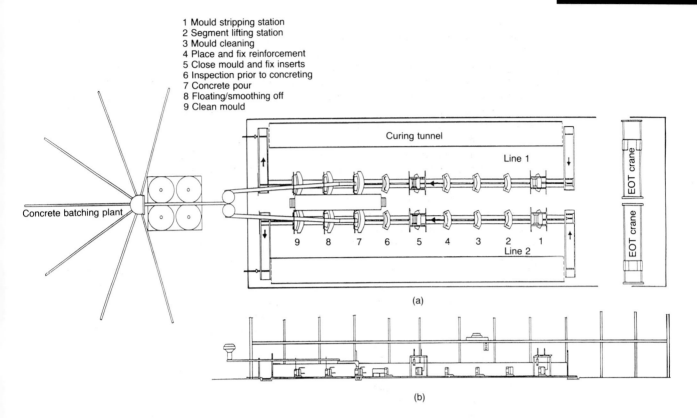

1 Mould stripping station
2 Segment lifting station
3 Mould cleaning
4 Place and fix reinforcement
5 Close mould and fix inserts
6 Inspection prior to concreting
7 Concrete pour
8 Floating/smoothing off
9 Clean mould

(a)

(b)

Fig. 17. Typical production building housing two production lines: (a) plan; (b) section

of 40 kN was required for the nodal welds, which was closely monitored by sample-testing the 35 000 ladder mats produced each week.

50. ROM were faced with the problem of meeting tolerances more associated with steel fabrication than producing large quantities of reinforcing cages. To achieve the fabrication tolerance of ±5 mm, specially developed precision-assembly jigs clamped the parts of the cage in position while they were tack-welded together using the MIG (metallic inert gas) welding process. The use of a fully welded cage was adopted to enhance the rigidity and inherent stability of the cage within the mould, while withstanding the needs of on-site handling and transportation.

51. To ensure dimensional compliance, all cages were checked in dummy moulds before being despatched to the production lines. ROM's initial reject/rework rate was of the order of 20%; however, continuous fine tuning of their process resulted in a reduction of the reject/rework rate to approximately 5% in the last year of production.

The production process

52. Various production methods were considered, ranging from simple conventional static moulds to highly sophisticated automated production systems. Having assessed the needs and received tenders for providing production equipment and moulds, the following system was developed.

53. Eight production lines were housed in four production buildings. Each line was made up of nine workstations, a curing tunnel and a conveyor system to bring concrete to the mould-filling station (see Fig. 17). The moulds travelled through the workstations at nominally 10 min intervals for stripping, cleaning, placing reinforcement, fitting inserts and concreting, before moving to the curing tunnel where they were steam-cured at 50°C for six hours before being lifted from the mould at 10 N/mm^2 (minimum) compressive strength, typically 15–20 N/mm^2. The overall time from the placing of concrete to the lifting of the segment from the mould was nominally 7 hours. The segments were cast on edge, to ensure that the extrados and grout pads were formed to the correct profiles and dimensions. Casting vertically had the added advantage of significantly reducing the total area of concrete to be hand-finished. Before leaving the building, each segment was covered by a special jacket to protect it against thermal shock and rapid moisture loss. The potential output of a production line was 144 per day.

54. In general there was a total of 35 No. different segment sizes/shapes (Table 2) and in each case there was a minimum of two different reinforcement configurations to accommodate the various ground conditions.

55. One of the major challenges was to find an acceptable method of holding the cages in the moulds; conventional concrete and plastic spacers were unacceptable because they provided a potential water path to the reinforce-

Table 2. Schedule of precast concrete tunnel linings

	No. rings to be made	No. segments in a ring	Segment wall thickness: mm	External dia: m	Nominal internal dia: m	Ring length m
Service tunnel						
Marine	13 200	6 plus 1 key	270	5·34	4·8	1·5
Land	4700	6 plus 1 key	410	5·62	4·8	1·5
Running tunnels						
Marine	26 500	8 plus 1 key	270/360	8·32	7·6	1·5
Land	9600	8 plus 1 key	540	8·68	7·6	1·5

ment. The problem was resolved by taking advantage of the very stiff concrete that positively held the reinforcement after placing/vibration, and a requirement by the client for two monitoring pins in each segment which protruded from the cage to the intrados surface.

56. The final solution was to hold the cage off the bottom of the mould with up to four specially manufactured concrete spacers, using the same materials as the segments. The bottom third of the cage was held by positively fixing the monitoring pins accurately to the reinforcement and mould, and the top of the reinforcement was fixed relative to the mould by temporary clamps which were removed after vibration. Close surveillance by the QC team demonstrated compliance throughout production.

Quality

57. At all stages during the precast yard development and operation, the management team recognized the immense importance of applying an effective quality system.[13] On site the Chief Engineer had the direct responsibility for establishing the system and operational procedures, along with managing a quality control/inspection team to monitor field application of the system.

58. The effectiveness of the self-inspection process adopted was reviewed daily by a resident Quality Assurance department, whose brief not only included a percentage review of the direct inspection activities, but also a close involvement in the approval and decision-making process. With an off-site reporting function, the QA department was responsible for the review/auditing of material suppliers, and for ensuring that the systems adopted were consistent with the overall policy. Each segment produced was given a detailed visual examination at the end of the production line. Segments were then either passed and transported to the storage area or quarantined on the grounds of a visual defect or a problem identi-

fied during the casting process. The quarantining of segments (typically 1–2% of production) rather than their immediate rejection had considerable benefits in terms of monitoring defect trends, which enabled management to target areas of the process for improvement.

59. Traditionally the industry has proved the dimensional compliance of tunnel segments by a combination of pre-production mould checks and elementary post-production dimensional and ring build checks. The specification for the PCC linings required the checking of some dimensions to ±0·1 mm which could not be demonstrated by adopting these methods. Following a detailed examination of the technology available, Translink decided to use a programmable co-ordinate measuring machine. Two Crown Windley series 3 CNC layout machines were purchased with Renishaw PMS300 controller and software packages. Development and verification of the complex 3D mathematical equations and segment measurement part-programmes were undertaken by the site engineering personnel at the Isle of Grain.

60. The success of the self-inspection quality system adopted is demonstrated by the fact that the rejection rate was 0·6% over the $3\frac{1}{2}$ year production period (latterly a typical weekly rate of 0·2% was achieved) with only 0·03% of segments rejected for a dimensional defect. Much credit must be given to Sacma for producing the moulds to tolerances generally 50% better than the concrete lining specification, and it should be noted that not one single mould had to be replaced during the manufacturing period. The maximum number of castings from any one mould was of the order of 1500.

Logistics

61. The logistics of manufacturing and transporting 442 755 precast concrete segments of 35 different types and sizes was a major planning exercise. At peak, three British Rail trains, each comprising 24 wagons and pulling a payload of approximately 2000 tonnes, completed the 180 mile round trip daily. Once the precast concrete linings had arrived at Shakespeare Cliff they were stored along with the SGI linings ready to be loaded onto the narrow gauge construction railway system, to be delivered to the respective workfaces.

Procurement of spheroidal graphite cast iron

Overview of requirements

62. Whereas it was a practical proposition to set up a precast concrete factory for the provision of the large number of concrete segments required, this was not possible for the SGI

rings. In 1986 the exact requirements for SGI linings were unknown. Early estimates showed a need for a flexible SGI lining for all tunnel diameters to cater for the heavier ground loadings, bolted cross-passage linings, smooth bore PRD linings, and bolted linings to cater for adverse ground conditions. Well over 30 lining types in all were required, each formed from a wide variety of precisely machined castings. These requirements totalled 102 000 tonnes over a period of 3 to 4 years. This was greater than the capacity of any single existing foundry producing tunnel linings.

Procurement strategy

63.   Initial tender enquiries to the major European suppliers for the total requirements indicated prices which were higher than those received pre-contract. This was due to the uncertainties of both design and actual quantities, and the major capital investments required to meet the order. It was therefore decided to divide the lining requirements into contracts which could be let in a phased sequence to suit the construction needs. This brought each package workscope within the capacity of existing foundries, allowed more competitive bids, and meant that the design could also proceed in a phased sequence.

64.   The fast-track nature of the project required contracts to be let on the basis of outline design drawings with the pricing structure allowing for 'extra over' items such as circumferential machining—with and without gasket grooves, grout holes and lifting points. Each package included a defined minimum quantity of rings and provisional quantities. Thus, competitive quotes were obtained which included sufficient pricing information to cover the developing design requirements.

65.   In order to reduce costs, a decision was taken to minimize the requirements for SGI linings, and the SGI/PCC hybrid rings were a most successful outcome of this policy. The final quantity of SGI linings was 38 133 tonnes, a reduction of 72% from the initial estimates.

Resulting subcontractors

66.   Following international pre-qualification procedures, enquiries were sent to all known lining manufacturers in Europe and the Far East. Orders were placed in the UK, France and Germany, and in each case the foundry had a particular technical advantage in machining technique, moulding size or capacity, which gave a commercial advantage for the particular package (see Tables 3–5 for details).

67.   Of particular note was the vacuum casting process used by Buderus to make the large complex segments for the hybrid rings, and the sophisticated use of computer-aided

Table 3. SGI lining packages and suppliers

Contract package	Supplier	Tonnes
AB5, AB6	Parkfield Foundries Ltd	593
SZ2, SZ3, SZ7	Stanton Plc	2693
SB2 (0)	Buderus BAS GmbH	724
R27X, R27X1, R36X, R36X1, R54X, R54X1	Buderus BAS GmbH	6430
R27P, R27P1, R36P, R36P1, R54P, R54P1	Buderus BAS GmbH	2787
SB1, SB2, SB3, AB5, AB7	Parkfield Foundries Ltd	7484
RB1, RB2, RB3	Parkfield Foundries Ltd	12 022
C1, C2, C3	Ferry Capitain	2921
AB8	Ferry Capitain	609
P1, TP1, P2, TP2	Ferry Capitain	1870
Total		38 133

machining used by Ferry Capitain to produce the smooth bore piston relief duct linings. On the first package for the conveyor tunnel rings, Parkfield significantly reduced the delivery period by using identical castings with an allowance for machining different radial joint angles. By changing the number of segments and the radial joint angles, different diameters could be produced using the same patterns.

68.   The supply of linings for adverse ground conditions was the most difficult area to manage. In the service tunnel the lining size had alternative uses in the cross-passages and pump stations. These later requirements were delivered early to provide a buffer stock against potential adverse ground requirements. This was not possible in the running tunnels because the TBMs could use more than the production capacity of any existing foundry, and in view of the difficulties encountered in the service tunnel an agreement was reached with Eurotunnel to provide financial assistance to the chosen lining supplier to enable him to substantially increase his manufacturing capacity. The assistance was paid in the form of a premium on 60 000 segments which would then be recovered on the next 50 000 segments, should these be required.

69.   Even with this 'insurance' the TBMs could still outstrip the foundry capacity. There was therefore a need to build up a contingency stock; the size of this contingency stock was again agreed with Eurotunnel. During the build-up of this stock, lining erection methods were developed in the tunnel such that PCC rings could be used in limited adverse ground conditions. However, the indications were that the situation in the running tunnel would be very much worse and that progress with SGI rings would be extremely slow, if not hazardous. It was therefore decided that for 700 m of each running tunnel the ground would be treated by grout injection from the service

Table 4. Schedule of SGI rings

Lining type type	Internal dia: m	Ring width: mm	Weight/ring: t	Segmentation O	T	K	Construction	Supplier	Total Ordered	Used	Locations
AB5	4·00	750	2·45	22	2	1	Machined/bolted	Parkfield	302	302	Conveyor tunnels
AB6	3·00	750	2·01				Machined/bolted	Parkfield	73	73	Conveyor tunnels
AB7	3·32	750	2·222	18	2	1	Machined/bolted	Parkfield	144	133	Cross-passages
AB8	4·00	750	1·450		2	1	Machined/bolted	Ferry Capitain	421	421	Cross-passages
SB1	4·93	750	2·116	12	2	1	Machined/bolted	Parkfield	596	596	Marine service tunnel adverse ground, pump stations, crossover, mid-channel junction, technical rooms, combined with SB 'O' opening sets
SB2	4·93	750	2·718	12	2	1	Machined/bolted	Parkfield	1414	1402	
SB3	5·29	750	4·742	12	2	1	Machined/bolted	Parkfield	283	283	Land service tunnel adverse ground, pump stations, technical rooms, breakthrough zone
RB1	7·86	750	4·164	16	2	1	Machined/bolted	Parkfield	1803	715	Marine running tunnels adverse ground, crossover approach, pump station, temporary crossover, mid-channel junction NVS shaft
RB2	7·86	750	5·879	16	2	1	Machined/bolted	Parkfield	450	306	
RB3	8·17	750	10·212	16	2	1	Machined/bolted	Parkfield	225	198	Land running tunnels adverse ground, opening construction, breakthrough zone, NVS shaft
C1	3·33	750	1·024	10	2	1	Machined/bolted	Ferry Capitain	344	343	Marine cross-passages
C2	3·33	750	1·165	10	2	1	Machined/bolted	Ferry Capitain	1208	1208	
C3	3·33	750	1·936	10	2	1	Machined/bolted	Ferry Capitain	600	600	Land cross-passages
P1	2·00	750	0·58	7	–	–	Tongue and groove/boltless	Ferry Capitain	1690	1690	Smooth bore marine piston relief ducts
TP1	2·00	N/A	0·492	7	–	–	Tongue and groove/boltless	Ferry Capitain	409	409	Smooth bore taper rings piston relief ducts
P2	2·00	750	0·73	7	–	–	Tongue and groove/boltless	Ferry Capitain	779	773	Smooth bore land piston relief ducts
TP2	2·00	N/A	0·625	7	–	–	Tongue and groove/boltless	Ferry Capitain	192	192	Smooth bore taper rings land piston relief ducts

Table 5. Schedule of SGI opening sets

Lining type	Internal dia: m	Set width: m	Weight set: t	Construction	Supplier	Total		Locations
						Ordered	Used	
SB2 (0)	4·93	6·0	14·48	Machined/bolted	Buderus	50	48	Marine service tunnel and pump station opening sets
SZ2 SZ3	5·38	6·0	19·0	Machined/flexible	Stanton	23 40	23 40	Marine service tunnel 6·0 m long opening sets for cross-passages and technical rooms
SZ7	5·46	6·0	33·0	Machined/flexible	Stanton	38	36	Land service tunnel 6·0 m long opening sets for cross-passages and technical rooms
R27X R27 X1	7·780	6·0 9·0	17·30 23·01	Machined/flexible SGI/PCC hybrid	Buderus	67 11	65 11	Marine running tunnel opening sets 6 and 9 m long for cross-passages and technical rooms
R36X R36 X1	7·600	6·0 9·0	23·74 29·93	Machined/flexible SGI/PCC hybrid	Buderus	93 8	90 8	Marine running tunnel opening sets 6 and 9 m long for cross-passages and technical rooms
R54X R54 X1	7·600	6·0 9·0	39·79 49·35	Machined/flexible SGI/PCC hybrid	Buderus	72 8	72 8	Land running tunnels for cross-passages and technical rooms
R27P R27 P1	7·780	3·0 6·0	6·39 12·16	Machined/flexible SGI/PCC hybrid	Buderus	68 30	68 30	Marine running tunnel opening sets 3 and 6 m long for piston relief ducts
R36P R36 P1	7·600	3·0 6·0	7·65 13·87	Machined/flexible SGI/PCC hybrid	Buderus	79 50	79 50	Marine running tunnel opening sets 3 and 6 m long for piston relief ducts
R54P R54 P1	7·600	3·0 6·0	14·15 26·89	Machined/flexible SGI/PCC hybrid	Buderus	62 30	62 30	Land running tunnel opening sets 3 and 6 m long for piston relief ducts

In all cases segmentation was designed to match equivalent PCC segments and required opening size.

Fig. 18. Mould being stripped prior to segment removal

Fig. 19. Technical room lined with SB3 SGI rings

process which sometimes over-ran the initial allowances. The exact numbers of opening sets, piston relief duct and cross-passage rings could not be finalized until the breakthrough zones were agreed and all other design aspects finalized, since the UK and French linings were not interchangeable. It is a great credit to the TML staff directly involved and the various foundries that, despite these problems, at no time was tunnelling hindered by lining shortages. It is also to their credit that the quantity of surplus linings was kept at such a low level (see Tables 4 and 5).

Conclusion

73. The nature of the contract resulted in the need for a close relationship between the designers, manufacturers and installers. The onerous durability and loading requirements, coupled with the need for fast erection and overall economy, strengthened this collaboration with the benefit of a better understanding of the various and somewhat conflicting needs of the three parties.

74. Changes were made to the design as the project proceeded, to suit fabrication and erection; these included

(a) detailing of reinforcing cages in the concrete segments;
(b) reducing the size of the grout pads on the extrados of the concrete segments to accommodate the hoods added to the TBM tailskins, when unexpectedly poor ground conditions were encountered;
(c) larger rebates in radial joints at the wedge keys to avoid damage during key insertion;
(d) provision for waterproofing gaskets in all cast iron segments.

75. Although the great success of the lining design and manufacture is well illustrated by the end results in terms of quality and progress, the Authors believe that the technical spin-offs will have a lasting widespread influence.

Acknowledgements

76. The Authors are grateful to C. Pollard who was Engineering Manager—Tunnel Construction, and was responsible for the procurement of SGI heavy duty cast iron linings. Also to J. Whorlow, who was the Deputy Construction Manager at the Isle of Grain, and was specifically responsible for quality control and the development of the segment measuring methodology. The Authors are also grateful to ET and TML for permission to publish this Paper.

tunnel.[14] The end result was that TML were left with over 1000 rings of running tunnel linings and the premium recovery applied only to a very small number of segments.

70. The hybrid SGI/PCC rings led to substantial time and cost savings; about 15 000 t of SGI were saved, valued at some £12 m. The fast erection cycle meant that about 16 weeks of TBM driving time were also saved, which gave 3 weeks' saving in overall project terms.

71. All SGI lining suppliers were required to comply with high QA standards. TML-appointed inspectors were placed in each foundry, and each was regularly visited by TML QA, Engineering and Expediting staff.

72. The fast-track aspects of the project inevitably led to delays in the procurement

References

1. GOULD H. B. *et al*. The design of the Channel Tunnel. *J. Instn Struct. Engrs*, 1975, **53**, 45–62; discussion *ibid* 537–542.

2. CURTIS D. J. *et al*. In situ ground and lining studies for Channel Tunnel. *Tunnelling '76*. Institution of Mining and Metallurgy, London, 1976, 231–242.

3. CRIGHTON G. S. and LEBLOND L. Tunnel design. *The Channel Tunnel*. Institution of Civil Engineers, London, 1989, 95–135.

4. NORIE E. H. and CURTIS D. J. The Channel Tunnel: Design for UK tunnels and related underground structures. *Strait Crossings*. A A Balkema, Rotterdam, 1990, 209–221.

5. CURTIS D. J. *et al*. The Channel Tunnel: design, fabrication and erection of precast concrete linings. *Tunnelling '91*. Institution of Mining and Metallurgy, London, 1991, 161–172.

6. HESTER J. C. *et al*. The Channel Tunnel UK/France—the UK TBM drive. *Rapid Excavation and Tunnelling Conference*, 1991.

7. O'ROURKE T. D. *Guideline for tunnel lining design*. The American Society of Civil Engineers, 1984.

8. CRIGHTON G. S. *et al*. Supplementary site investigations for the Channel Tunnel 1986–7: an overview. *Tunnelling '88*. Institution of Mining and Metallurgy, London, 1988, 55–77.

9. CURTIS D. J. Discussion on MUIR WOOD A. M. The circular tunnel in elastic ground. *Geotechnique*, 1975, **25**, No. 1, 114–127.

10. Recommendations for the use of convergence–confinement method. Draft report of AFTES Working Group No. 7. *Tunnels et Ouvrages Soutèrrains*, 1986, No. 73.

11. WOOD J. G. M. *et al*. Improved testing for chloride ingress resistance of concretes and relation of results to calculated behaviour. *3rd International Conference on Deterioration of Reinforced Concrete in the Arabian Gulf*, 1989.

12. GREATOREX C. B. and BATEMAN T. New development in SGI tunnel lining. *ICE Symposium on Tunnel Construction*. Public Works Congress: 11(4).

13. EVES R. C. W. and PARTRIDGE R. Quality Assurance at Grain. *World Tunnelling*, 1990, Aug., 268–271.

14. CRAWLEY J. D. and POLLARD C. Ground treatment to improve tunnel progress on the Channel Tunnel marine drives. *Ground Engineering*, 1992, Jan., 27.